ROADSIDE GEOLOGY

of
Hawai'i

Richard W. Hazlett
Donald W. Hyndman

1996
Mountain Press Publishing Company
Missoula, Montana

PRINTED IN THE UNITED STATES OF AMERICA

Unless otherwise indicated, all photographs in this book are by the authors.

Library of Congress Cataloging-in-Publication Data

Hazlett, Richard W.
  Roadside geology of Hawai'i / Richard W. Hazlett, Donald W. Hyndman.
    p. cm.
  Includes bibliographical references (p. -) and index.
  ISBN 0-87842-344-3 (alk. paper)
  1. Geology—Hawaii—Guidebooks. 2. Hawaii—Guidebooks.
I. Hyndman, Donald W. II. Title.
QE349.H3H39 1996
559.69—dc20   96-31763
                                                        CIP

Mountain Press Publishing Company
P.O. Box 2399 • Missoula, MT 59806
(406) 728-1900 • (800) 234-5308

To the memory of
Harold T. Stearns, 1900–1986,
pioneering Hawai'i geologist.

Contents

Preface

We wrote this book for people who would like to learn about the geology of Hawai'i. Most of what is known about Hawaiian geology is widely scattered through the professional literature, much of which is difficult to locate and even more difficult to read. Because almost all published geologic research is buried in governmental and university libraries, the results should be made available to the general public.

The book begins with a general introduction to the geology of Hawai'i. A chapter covers each of the inhabited and easily accessible main islands. Each chapter begins with a general discussion of the rocks of that island, then proceeds with a series of road guides that provide the local details.

Geologists sometimes use more than their fair share of jargon, which makes what they do and say seem deeper and more mysterious than it really is. Most of the rocks and ideas that geologists investigate are understandable in the everyday vernacular. We did our best to eliminate as much jargon as we could, and we believe we got rid of most of it. But some words, especially the names of rocks, have no vernacular equivalents, and you cannot write about geology without discussing rocks. A glossary is provided at the back.

A Note on Hawaiian Language

The following guidelines will assist readers who are unfamiliar with how to pronounce Hawaiian words. The Hawaiian language is characterized by clusters of vowel combinations and a small number of consonants. Two diacritical marks used in Hawaiian affect pronunciation. They also affect a word's meaning—for example, lanai (stiff-backed) and lānai (porch, verandah), and ono (a type of fish) and 'ono (delicious).

Hawaiian has eight consonants (h, k, l, m, n, p, w, '). The first seven are pronounced roughly the same as in English. The sound represented by an okina, a single opening quotation mark ('), is a glottal stop, pronounced like the sound in "oh-oh."

When you know how to pronounce Hawaiian vowels, pronouncing Hawaiian words is not difficult. Two vowels, *e* and *i*, are pronounced differently than they usually are in English: *e* sounds close to a long *a*, and *i* sounds like a long *e*. Vowels with a *kahako* or macron—a bar—over them are pronounced with greater than normal stress and duration. Pronounce Hawaiian vowels as follows:

a as in l*a*va	m*a*h*a*lo
e as in b*e*t	h*e*l*e* (go)
ē as in h*ay*	nēnē (Hawaiian bird)
i as in f*ee*t	Waik*i*k*i* (Wye-kee-kee)
o as in h*o*le	al*o*ha
u as in b*oo*t	h*u*la

Dipthongs produce various vowel sounds: *au* as in Maui, *oi* as in *poi*. A long *i* sound comes from the dipthong *ai,* as in *makai* (toward the sea).

Determining where to place stress in a Hawaiian word can be baffling for someone not used to hearing Hawaiian words pronounced. For shorter words, the next-to-last syllable is often stressed (wa*hi*ne, al*o*ha, O*'a*hu). Also, a syllable with a long vowel or a dipthong—lā (sun), kai (sea)—is often stressed.

Two Hawaiian words you will encounter frequently in this book are descriptive terms for two different kinds of lava: *'a'ā* (ah-*ah*), sharp, rough, cindery lava, and *pāhoehoe* (*pah*-hoi-hoi), smooth, ropy lava.

Acknowledgments

We thank the following people for their help in producing this book. David Alt of the University of Montana contributed much to the book's content and, with Jane Taylor, Kathleen Ort, and Jennifer Carey, improved its readability. James G. Moore of the U.S. Geological Survey provided specific information on the giant tsunami deposits on Moloka'i and Lana'i. Ronald Greeley of Arizona State University provided a map of Kaūmana lava tube. Byron Kesler of Volcano, Hawai'i, did much of the drafting. Jean MacKay of Pomona College helped with the typing. Gerhard Ott of Pomona College developed many of the photographs taken by Pomona students David Saltzer and Ronald K. Gebhardt. Drafts were reviewed by Thomas L. Wright, Robin Holcomb, Christina Heliker, and Robert W. Decker, all research geologists with the U.S. Geological Survey; and by John Sinton of the geology department at the University of Hawai'i.

Mahalo nui loa!
(Thank you all very much!)

1
Geology in the Hawaiian Islands

For geologists accustomed to working on continents, Hawaiʻi is a strange and remote land. It sits near the center of the world's largest ocean basin, thousands of miles from any continent or tectonic plate boundary, as far removed in the tale it tells as the surface of another planet.

The geology of Hawaiʻi seems basically simple at first glance: The islands are a group of volcanoes in the ocean. Closer examination reveals much more. Our scientific understanding of the geologic evolution of Hawaiʻi is still developing. Tantalizing mysteries remain. We present this book in the spirit of a progress report.

Of Time and the Hot Spot

Time is the essence of geology, in Hawaiʻi or anywhere. Time means something quite different to a rock than to a human. To us, a century is a long time; to a rock, a century is nothing.

Dinosaurs flourished 75 million years ago. No polar ice caps existed then. Sea level was so high that most of the North American midcontinent was flooded. The Pacific Ocean was much larger. Far out in the depths of this wide ocean, early versions of the Hawaiian Islands had already appeared, close to where the Hawaiian Islands are now. Since then, these islands have continued to appear and then sink beneath the waves, carried off to the northwest by the drifting ocean floor.

The geologic explanation for the Hawaiian Islands postulates a small area of abnormally hot rock, a "hot spot" deep beneath the ocean floor. The hot spot is firmly rooted in the earth's interior, while the bedrock of the ocean floor glides slowly over it. Molten rock erupts from the hot spot onto the ocean floor and builds a volcano, which grows above the waves until the ocean floor carries it past the hot spot. Beyond its lava supply, the old volcano goes out of business, and a new volcano begins to build above the hot spot.

This process created a chain of volcanoes extending westward from the Big Island to just past Kure Atoll, then north to the western end of the Aleutian Islands. The Big Island still lies mostly over the hot spot. At this end of the chain, five volcanoes have erupted in historic time. Elsewhere in the main Hawaiian archipelago, the volcanoes are extinct, but still young enough to

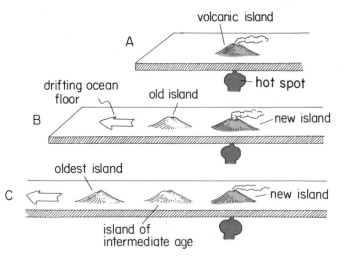

volcanic island

A

drifting ocean floor

old island

B

new island

hot spot

The Hawaiian Islands sit on top of a nearly stationary hot spot. Each island forms as it passes over the hot spot and then moves off it with the drifting ocean floor.

oldest island

C

new island

island of intermediate age

form high-standing islands. Between the main islands and Kure Atoll, the volcanoes, also extinct, have sunk to form atolls and reefs. Beyond Kure, the ancient volcanic summits of the Emperor Seamounts have submerged beneath the waves.

What is a hot spot, and is it really stationary? Why does the ocean floor move across it? Why did it produce a chain of separated islands rather than a single long volcanic ridge? These questions baffle and delight geologists. The more we try to answer them, the more questions we raise.

Many kinds of evidence show that the earth's interior is very hot: We see molten lava pouring from erupting volcanoes; geysers jet hot water into the sky; and miners who work deep underground find rocks there hot to the touch. Geophysicists estimate that the core of the earth may have a temperature of almost 10,000 degrees Fahrenheit.

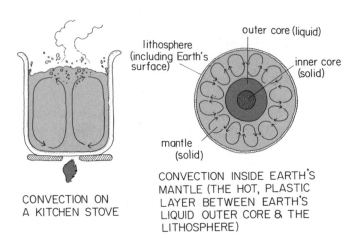

lithosphere (including Earth's surface)

outer core (liquid)

inner core (solid)

mantle (solid)

CONVECTION ON A KITCHEN STOVE

CONVECTION INSIDE EARTH'S MANTLE (THE HOT, PLASTIC LAYER BETWEEN EARTH'S LIQUID OUTER CORE & THE LITHOSPHERE)

A fluid heated at its base rises and circulates.

Radioactive decay of certain elements furnishes most of the earth's internal heat supply. It drives a very slow convection inside the earth. Extremely hot rock rises, loses heat through volcanoes, then sinks back into the depths. You can see the same process in a pot of thick soup slowly stirring itself as it simmers on the kitchen stove. In both cases, heat and gravity drive the movement.

We all find it difficult to imagine solid rock flowing in convection currents inside the earth. Rocks at the earth's surface are cold, stiff, and brittle. But under the high temperatures and pressures in the earth's interior, rocks become soft and plastic in much the same way that hard wax becomes soft and pliable as it warms in your hand.

Hot rock in the earth sometimes rises in narrow plumes, like smoke rising from a chimney. A plume heats the underside of the earth's outer rind, the

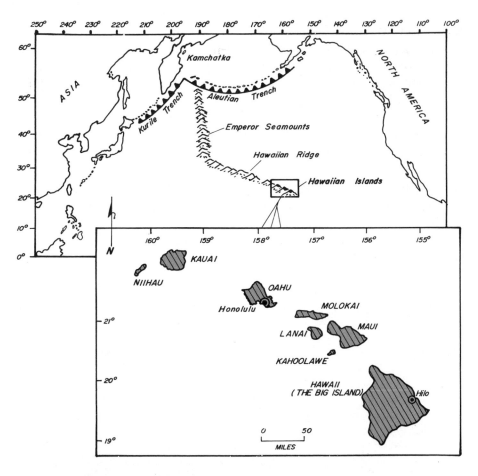

The Hawaiian Islands and the Hawaiian–Emperor Seamount chain.

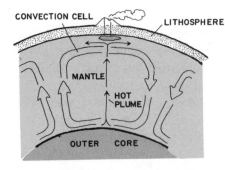

Heated rock flows beneath a hot spot. The long column of rising rock is called a plume.

lithosphere, the way a welder's torch burns through a sheet of boiler plate. It melts the rock, which erupts through volcanoes above the hot spot. Having lost heat through the volcanoes, the plume rock then spreads beneath the moving lithosphere and sinks back into the depths.

Hot spots are undetectable to most of the conventional methods used to examine the earth's interior. Geologists have no firm data to help explain how hot spot plumes start rising, or from what depth.

The vapors blowing out of Hawaiian volcanoes contain surprisingly high concentrations of the rare elements iridium and osmium. Some geologists regard this as evidence that the plume is rising from an extremely deep root. Others do not.

Some geologists believe that hot spot plumes start spontaneously where large masses of hot rock rise from peaks in the boundary at the top of the earth's molten core. Others suggest that they originate at depths of only 400 to 450 miles, where pressure forces the abundant mineral olivine to change its crystalline structure, forming a boundary from which escaping heat can rise. Some geologists speculate that plumes develop at sites where enormous asteroids strike the earth and explode, opening a path for heat to escape from the interior. A few believe that Bowers Ridge, a large, curving structure on the floor of the Bering Sea, could be the site of an asteroid impact and the starting point of the Hawaiian hot spot.

Whatever the origin of hot spots, geologists have identified 40 to 50 active ones. Hawai'i, the Yellowstone Volcano, and Iceland are among the most active and most familiar.

Possible motion of heated rock inside the earth. The mantle moves in layers, and hot-spot plumes may originate on the horizontal boundary separating them.

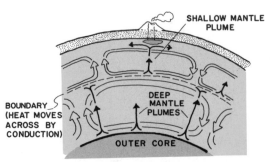

The Mantle and the Crust

The outer rocky layer of the earth is called the crust. The boundary between it and the mantle beneath is so sharp that earthquake waves reflect from it and echo back to the surface. Geologists measure the thickness of the crust by timing the echo.

The continental crust is between 15 and 40 miles thick, made mostly of granite, schist, and gneiss, pale rocks consisting largely of the minerals feldspar and quartz. The Hawaiian Islands have no continental crust: They are strictly oceanic.

Oceanic crust consists almost entirely of basalt, a black volcanic rock with sparse, scattered crystals of the black mineral pyroxene, pale gray plagioclase feldspar, and apple green olivine. Oceanic crust is generally about 5 miles thick, but much denser than continental crust, which explains why the ocean floor floats lower than the continents. The average depth to the ocean floor in the Hawaiian region is about 16,000 feet.

Beneath the crust lies the mantle, reaching all the way down to the earth's core. The shallow mantle is mainly a rock called peridotite, which consists chiefly of pyroxene and olivine. The mantle beneath Hawai'i is a variety of peridotite called lherzolite, mainly pale green olivine and two varieties of pyroxene, one dark green and the other black.

Hawaiian lherzolites contain small amounts of other minerals. Spinel, an oxide of magnesium and aluminum, forms tiny octahedral crystals of various colors in lherzolite that comes from depths of less than 30 miles. Lherzolite from greater depths contains beautiful, dark red garnets. Laboratory experimental work shows that the molten basalt magma rising from the Hawaiian hot spot forms through partial melting of garnet lherzolite at depths between 40 and 60 miles.

Lithosphere and Asthenosphere

Even though the crust and uppermost mantle consist of distinctly different rocks, their mechanical behavior is essentially the same. Therefore, geologists generally refer to the two layers together as the lithosphere; it is the earth's outer rind.

Beneath the lithosphere lies the asthenosphere, where the mantle rock flows as though it were soft wax. The temperature there is probably just high enough to melt a small portion of the rock, despite the high pressure. The asthenosphere is a greased skid, a slippery banana peel beneath the moving lithosphere. The lithosphere is a mosaic of pieces called plates that fit together like the bones in a skull. All the plates move, each in its own direction, in what does not appear to be part of an orderly pattern. Plates slide past each other at some boundaries, collide at others, and pull away from each other at the rest. The Hawaiian Islands ride on the Pacific plate, which is carrying them northwest at about 4 inches every year.

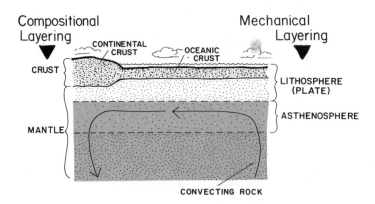

Compositional Layering ▼

CONTINENTAL CRUST — OCEANIC CRUST

CRUST

MANTLE

Mechanical Layering ▼

LITHOSPHERE (PLATE)

ASTHENOSPHERE

CONVECTING ROCK

In speaking of the earth's composition, geologists refer to the crust and mantle. In speaking of the earth's mechanical behavior, they refer to the lithosphere and asthenosphere, which do not correspond to the compositional layers.

Where two plates collide, the heavier one sinks into the earth's mantle. Oceanic trenches are places where a sinking plate bends down as it begins its long plunge into the depths. The Pacific plate is sinking into trenches all around its northern and western perimeter. This has already carried parts of the oldest Hawaiian hot spot volcanoes into the Aleutian trench, and down into the mantle. If current trends continue, the modern Hawaiian Islands will sink through the Kurile trench off the northeast coast of Japan in another 40 million years.

As oceanic trenches gobble up the ocean floor, new oceanic crust is forming elsewhere. That takes place at oceanic ridges, where two plates pull away from each other. Molten basalt erupts through the gap opening between them

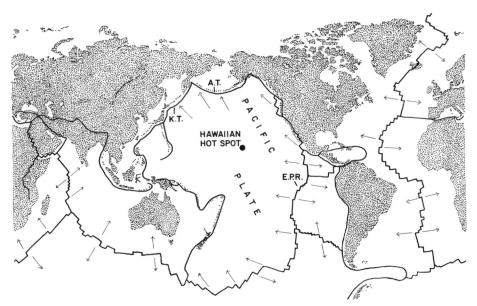

The earth's major plates. Arrows indicate the direction of their movement. New oceanic crust is forming at the East Pacific Rise (E.P.R.), and old oceanic crust is sinking through the Kurile and Aleutian trenches (K.T. and A.T.).

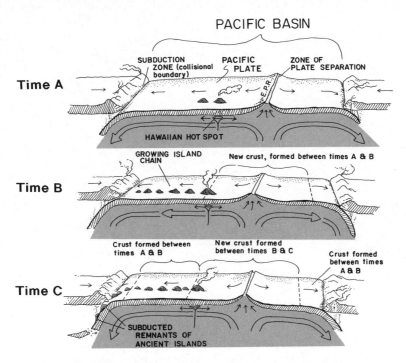

The floor of the Pacific Ocean sinks around most of its margins, but new oceanic crust forms in the basin along the East Pacific Rise, an oceanic ridge.

to make new oceanic crust. Some geologists have compared this process to what would happen if two patches of ice on a lake were slowly moving apart with the water freezing as it welled up between them. Both pieces of ice would grow, but the gap between them would remain at a constant width.

Molten Rock

Underground molten rock is called magma; if it erupts, we call it lava. Some lava erupts quietly in the form of lava flows; the rest erupts explosively to make pyroclastic debris: bombs, cinders, and ash. All are volcanic rocks and typically consist of microscopic crystals, perhaps with a few large enough to be seen easily.

Some magma crystallizes and cools underground, forming intrusive rocks. These look quite different from volcanic rocks, although their compositions are the same. Intrusive rocks typically consist of crystal grains large enough to be seen without a magnifier.

As much as 40 percent of the garnet lherzolite in the area of the Hawaiian hot spot melts into basalt magma. The molten magma erupts within an area encompassing the three vigorously active volcanoes in Hawai'i: Kīlauea, Mauna Loa, and the Lo'ihi Seamount, which has yet to grow above sea level. The greatest distance between those volcanic summits is about 50 miles, which approximates the diameter of the Hawaiian hot spot.

Basalt with vesicles, small cavities that form when gas bubbles are trapped in cooling lava.

One or 2 percent of the weight of the magma is gaseous, mainly steam, carbon dioxide, and sulfur dioxide. These gases enter the magma at great depth, where it is under high pressure. As the magma nears the surface, where the pressure is lower, the volcanic gases bubble out of solution the way gas bubbles out of uncorked soda pop. Escaping gases froth the magma, helping make it light enough to rise and causing it to fountain or explode when it reaches the surface. Most of the volcanic gases blow off during eruptions and do not become part of the volcanic rock. Gas trapped in the lava creates small bubbles called vesicles.

Magma rises because it is lighter than the rocks it rises through. Detailed studies of the earthquakes generated by rising magma show that a fresh batch of Hawaiian magma can erupt within a few weeks or months after it begins rising in the mantle. The molten rock normally pauses on the way up. Geologists call a reservoir of stored, waiting magma a magma chamber.

The most pronounced pause is a mile and a half or more below the surface, where the magma apparently loses buoyancy when its density matches the surrounding rock. A large volume of magma may collect in a magma chamber at that depth. The trigger for eruption may be the ascent of new magma into the chamber from below. As the chamber swells, its roof cracks, allowing gases at the top of the magma to expand and propel the molten rock to the surface.

Mauna Loa erupts such huge amounts of lava that geologists are convinced it must be tapping an enormous chamber of stored magma. They use instruments called tiltmeters to monitor the ebb and flow of molten rock through the magma chamber. Tiltmeters are extremely sensitive in measuring the slope of the ground. They show when the flanks of the volcano be-

8

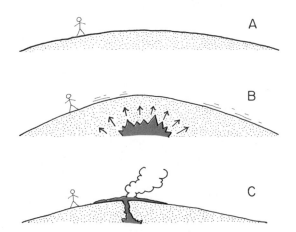

A volcano swells with magma before an eruption, then deflates during the eruption.

come steeper as a magma chamber fills, then flatten as the magma erupts. It is almost as though the volcano were inhaling, then exhaling.

Erosion has exposed the interiors of extinct volcanoes on the older Hawaiian islands, so we can see how they grew. As expected, they consist largely of lava flows with lesser volumes of pyroclastic debris. They also contain intrusive rocks, mainly dikes, and some basalt sills. A dike is a steep fracture filled with intrusive basalt; a sill is a more or less horizontal layer of basalt intruded between the lava flows. Dikes and sills are especially abundant in volcano summits, but many dikes also cut across the flanks in narrow bands called rift zones. In some places, erosion exposes masses of intrusive rocks, old magma chambers that completely crystallized. Many of these are more or less globular masses of gabbro called bosses. Gabbro is chemically identical to basalt and consists of the same minerals, but in larger crystals.

Fissures are the most common type of volcanic vent in Hawaiian eruptions. They open ahead of rising magma, then break the surface as long cracks. First, super hot volcanic gas and steam jet from a new fissure, spraying out shreds and globs of molten magma. Then, as the fissure widens, a torrent of lava pours out and may spread for miles downslope. After the eruption ends,

Dikes cut across volcanic layers; sills sandwich between them horizontally.

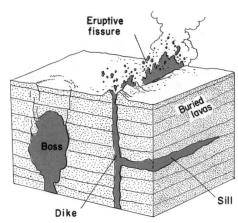

9

The opening of an eruptive fissure as a dike breaches the surface near the southern rim of Kīlauea caldera. July 1974.

A fissure at Kīlauea partly filled with lava that drained back into it in July 1974.

magma continues to fill the original fissure and crystallizes, becoming a dike. Most dikes exposed by erosion of older volcanoes are probably what remains of the natural plumbing that fed magma into fissure eruptions.

Some Hawaiian eruptions do not follow the typical pattern. Instead, unusually high concentrations of gas and steam may blast most of the magma out as black clouds of dusty ash or as heavier charcoal-sized cinder lumps. The cinders fall near the vent, building a cone, while the ash may drift far downwind. Cinders and ash are two examples of pyroclastic debris, a catchall term volcano specialists use to describe bits of lava exploded from a vent.

Shield Volcanoes

Hawaiian magmas are among the hottest on earth. Temperatures as high as 2,200 degrees Fahrenheit have been measured in molten lava at Kīlauea. Other kinds of lava erupt from some mainland volcanoes at temperatures as low as 1,560 degrees.

The type of eruption and type of volcano depend on the composition of the magma, the temperature of the magma, and how much gas it contains. The high temperature of Hawaiian basalt magma, its chemical composition, and its gas content all favor eruption of very fluid lava that can travel for miles and spread out in flows rarely more than 30 feet thick.

A long series of such eruptions builds a gently sloping mound of thin lava flows around the vent. If such a mound is only a few miles across, geologists call it a lava shield. Much larger ones, tens of miles in diameter, are called shield volcanoes.

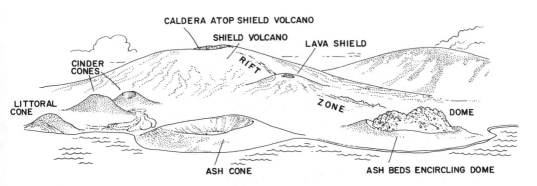

Basic types of volcanic landforms in the Hawaiian Islands.

Cinder cones dot the top of Mauna Kea (foreground). Mauna Loa, in the background, is one of the the world's largest shield volcanoes.

Mauna Loa and Kīlauea on the Big Island are shield volcanoes on a gigantic scale. So gentle are their slopes, so broadly arching their summits, that visitors accustomed to the steep cones of mainland volcanoes find it difficult to grasp what they are seeing. Smaller lava shields that grew during single eruptions on the flanks of Mauna Loa and Kīlauea include Mauna Iki and Mauna Ulu, in Hawaii Volcanoes National Park.

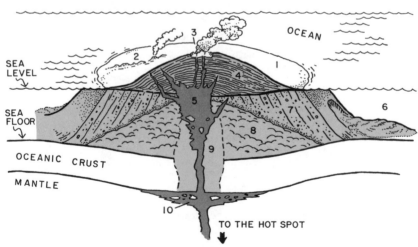

Section through a typical Hawaiian shield volcano.

1, *Volcano;* 2, *rift zone;* 3, *summit caldera;* 4, *lava flows above sea level;*
5, *magma chamber;* 6, *submarine slide;* 7, *basalt rubble;* 8, *pillow basalts;*
9, *gabbro; and* 10, *magma chamber.*

The lava shield of Mauna Ulu grew in the east rift zone of Kīlauea between 1969 and 1974. —U.S. Geological Survey photo

While the shield volcano stands above the hot spot it grows rapidly, perhaps erupting continuously for years. The volcano may pour out a cubic mile or more of fresh lava onto the slopes every century. The typical rock erupted during this early phase of activity is an extremely common type of basalt called tholeiite.

Like most other basalts, tholeiite is black, or nearly black; it commonly contains small, scattered crystals of yellowish green olivine. During the main stage of shield growth, tholeiite basalt tends to erupt frequently and gently, at least by volcanic standards. You can watch most of these eruptions at remarkably close range.

Dying Volcanoes

Eruptions become infrequent as the shield volcano starts the long journey away from the hot spot. Volcanic activity continues in sporadic, widely scattered eruptions around the summit, leaving the upper slopes of the volcano littered with cinder cones and rough lava flows. Meanwhile, weathering and streams convert the original volcanic landscape on the lower slopes of the mountain into an erosional landscape.

The big volcanoes stop erupting tholeiitic basalt almost as soon as they begin to decline. A new type of lava appears, enriched mainly in the alkali element sodium. This lava comes in many varieties, with such names as hawaiite, mugearite, and ankaramite.

It is difficult, and in many cases impossible, to distinguish these lava types in the field. You generally need a chemical analysis, a good view of a thin section of the rock through a specialized petrographic microscope, and a lot of experience with basalt. Think of them all as alkalic basalt.

13

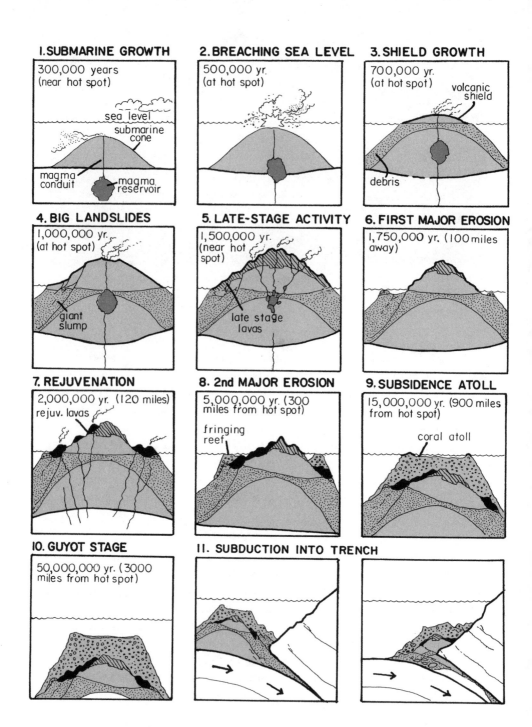

Typically, a Hawaiian volcano begins on the ocean floor, matures into an island shield volcano, then declines as the moving plate carries the island off the hot spot.

Deep erosion of the extinct shield volcano of Kaua'i reveals stacks of lava flows once buried deep within the mountain. This view into Waimea Canyon shows why it is sometimes called the Grand Canyon of the Pacific.

The vents that erupt alkalic basalts in the late stage of volcanism are near those that erupted olivine tholeiite during the main stage of activity. So it seems the late-stage alkalic magmas must follow the volcanic plumbing established during the preceding stage of shield growth.

The waning activity during the late stage leaves the upper slopes of the volcano littered with widely scattered cinder cones and lava flows. The big volcanoes stop erupting tholeiitic basalt as soon as they begin to decline. Their late-stage activity produces a variety of lavas, mainly basalt enriched in the alkali element sodium. These basalts come in many varieties with such names as hawaiite, mugearite, ankaramite, and nephelinite. They all look very much like ordinary basalt. Think of them as alkalic basalt.

Years of laboratory research and intellectual gymnastics have not enabled geologists to imagine how tholeiitic basalt magma could turn into alkalic basalt. Nothing that could plausibly take place in a magma chamber would cause that change. Alkalic basalt magma must melt with that composition in the mantle. Geologists can imagine several more or less plausible scenarios to explain how it might happen. Most theories involve melting a smaller proportion of the lherzolite in the mantle, or melting it at depths greater than those at which tholeiitic basalt magma can melt.

Kohelepelepe (Koko Crater) is a pyroclastic cone, the largest of the rejuvenated volcanoes in the Koko Rift at the southeastern tip of Oʻahu. The roadcut in the foreground exposes layers of fragmental debris erupted from it. —David Saltzer photo

Some of the late vents also erupt a lava called trachyte, consisting largely of feldspar, which is very rich in sodium. You can recognize trachyte by its pale gray color, which contrasts notably with that of ordinary black basalt. Many geologists believe trachyte magma develops during long storage in a slowly crystallizing alkalic basalt magma chamber, where the denser crystals of pyroxene and olivine settle, and the lighter crystals of feldspar float or stay behind. The result would be a layered magma chamber with trachyte at the top and basalt below.

The patient processes of erosion carve the quiet volcano into rugged terrain for tens of thousands of years after the last of the late eruptions. Then, on many aging Hawaiian islands, another round of volcanic activity begins. Geologists call this the rejuvenated stage of volcanism.

Eruptions in the rejuvenated stage appear to be unrelated to earlier eruptions. Individual vents are widely scattered across the volcano, and may lie in areas that never erupted. Many are near the coast; some are on the sea bed far offshore. Each seems to have its own plumbing to the lava source in the mantle. Evidently, the last batches of magma rise along new channels, perhaps because the old shield plumbing system has clogged. Eruptions in the rejuvenated stage create small lava shields, cinder and ash cones, and lava flows.

Rocks erupted during the rejuvenated stage of volcanic activity include some alkalic basalts like those that erupt during late-stage activity, along with others, called nephelinites, which are much more strongly enriched in sodium. Those may contain unusual minerals, like nepheline and melilite,

Air view of the east rift zone on Kīlauea, with an eruption in progress at Puʻu ʻŌʻō. Mauna Loa is in the background.
—U.S. Geological Survey, Jim Griggs photo

which resemble feldspar but contain larger amounts of sodium and smaller amounts of silicon.

Xenoliths

Any older fragment of rock incorporated in a lava flow or ash bed is called a xenolith. Many of them come from inside the volcano; some come from the mantle. Xenoliths from the mantle are found only in lavas erupted from volcanoes in the rejuvenated stage, never in the lavas erupted earlier. Many geologists believe the heavier xenoliths settle out of the magma chambers beneath and within the big shield volcanoes, as sediment settles in a lake. If so, then the xenoliths in the lavas erupted during rejuvenated activity may tell us that magma rises without pausing in magma chambers on the way up from the mantle.

Rift Zones

Immense mountains, these Hawaiian volcanoes. They rise from ocean floor as deep as 18,000 feet to a maximum elevation above sea level of 13,796 feet on Mauna Kea, a total of more than 31,000 feet. Hawaiian volcanoes are essentially huge piles of weak rock slowly spreading as they settle under their own weight. The spreading gradually opens narrow bands of fractures that permit magma to intrude far into the shield flanks, sometimes resulting in fissure eruptions tens of miles from a volcanic summit. Geologists call these fractured eruptive belts volcanic rift zones.

From the air, a rift zone looks like a long line of pyroclastic cones and lava shields trending straight down the slopes of the volcano. Because frequent

Gigantic fissures in the southwest rift zone of Kīlauea, where it meets Crater Rim Drive. These cracks opened as dikes intruded in 1868. Lava erupted from nearby fissures partially filled them in 1971.

eruptions rapidly build up the rift zone, the flanks on either side of a rift zone typically are steeper than the slope along the line of the rift zone itself.

Most Hawaiian volcanoes have three prominent rift zones radiating like spokes from their summits. Where a young shield builds up against the slope of an older volcano, only two rift zones develop. They separate the shield's stable side, next to the older volcano, from the side that slopes unsupported to the deep ocean floor. Kīlauea is an good example; an older neighbor, Mauna Loa, buttressed its northern flank and determined the orientation of its two rift zones.

Pyroclastic Cones

During the main stage of volcanism, rift zone eruptions produce fissures and lava shields. Later, cinder cones appear. Cinder cones normally form during a single eruption lasting a few weeks or months, and they rarely grow to be higher than a thousand feet. Craters shaped like cups truncate their tops. Toward the close of activity, as the last gas escapes from the magma, the cone stops growing, and a lava flow may burst through the loose cinder at the base. The lava erupts there because the loose pile of cinder has no internal strength, the same way milk runs out at the base of a pile of corn flakes. The flow may raft off a whole section of the cinder cone as it pours away, leaving behind a horseshoe shaped rim.

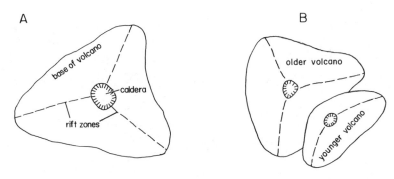

A, *The typical arrangement of rift zones around a single shield volcano and* B, *in a pair of volcanoes where one partially supports the other.*

Once it has gone out of business, a cinder cone normally never erupts again; most are single-shot eruptive features. The next outburst in the neighborhood will build a new cinder cone somewhere nearby.

In many cases, steam filters through the cinder cone for months or even years after the eruption has ended. The steam oxidizes the iron in the cinder, staining the cinder cone red with iron oxide. The result is a red cinder cone, a source of red road and garden gravel.

Lava bombs and blocks abound on cinder cones. Bombs are blobs of lava blasted out of vents. They become streamlined by the wind as they sail through the air. Blocks are jagged chunks of older rock that rushing volcanic gases rip from the throat of the volcano and throw out without streamlining. Bombs and blocks up to several feet in diameter are fairly common. Where they land on soft beds of ash or cinder, the layers sag under their weight.

Late-stage cinder cones dot the summit of the Mauna Kea volcanic shield.

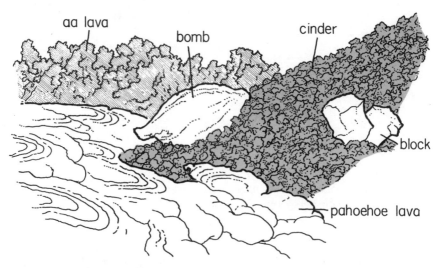

Major types of volcanic deposits in Hawai'i. Pāhoehoe and 'a'ā are two different kinds of lava from surface flows, one rough and the other smooth. Cinders, bombs, and blocks are explosively ejected lava fragments.

Steam is mainly what makes some volcanic eruptions explosive. Magma is likely to encounter shallow underground water if it rises into coastal areas during the rejuvenated stage of activity. The water boils into steam, which explodes in huge blasts, ejecting magma as a fine, hot mist mixed with chunks of older rock. The tiny mist particles are volcanic ash. They accumulate around the vent as an ash cone that is considerably broader than a cinder cone, generally with a shallow crater. Lē'ahi (Diamond Head) and some of its neighbors near Honolulu are good examples of ash cones.

Escaping gases sometimes inflate magma into a spongy froth called pumice, which is comparable to the head on a glass of beer. Many people recognize pumice as a porous, gray scrubbing stone so light that it floats, or almost floats. Hawaiian pumice differs from the mainland variety in being golden and very porous. Hawaiian pumice has so many vesicles, or gas bubbles, that they quickly connect; the pumice becomes waterlogged and sinks.

Volcanic pumice cones usually form downwind of a vent and may grow several hundred feet high. Mixed lava and pumice cones may grow even higher.

Where ash or pumice cones develop from eruption of trachyte magma, a dome may form. A dome is a huge lava mound that grows from a vent after the magma has blown off most of its gas, allowing the remaining molten rock to ooze out. Domes can completely fill the crater of a pyroclastic cone. While not common in Hawai'i, trachyte domes do make a few prominent landmarks.

Another kind of cone sometimes forms when molten basalt lava pours into the ocean, exploding into billowing clouds of steam and debris. The

A volcanic bomb about 2 feet across on Haleakalā, Maui.

Volcanic blocks strewn across a slope inside the basin atop Haleakalā, Maui.

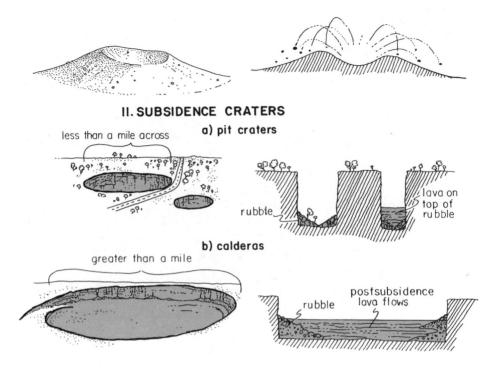

I. EXPLOSION CRATERS

II. SUBSIDENCE CRATERS
a) pit craters

less than a mile across

lava on top of rubble

rubble

b) calderas

greater than a mile

postsubsidence lava flows

rubble

Types of Hawaiian craters.

expanding steam can rip the lava into millions of black, glassy cinders, which pile up in mounds called littoral cones. Unlike other types of pyroclastic cones, these lie far from the eruptive vent, at the end of a lava flow. Many littoral cones lack well-defined craters; waves rapidly wash them away.

Craters

Volcanic craters come in many varieties, from different origins. Some open explosively during pyroclastic eruptions; others form as the volcanic surface quietly subsides into an emptying magma chamber. Most craters are distinctive enough that you can recognize their type and know how they formed.

Explosion craters, the kind that form during pyroclastic eruptions, are generally shaped like a cup sitting on top of a pile of pyroclastic debris. Craters on cinder cones are rarely more than a few hundred feet across; those on ash cones may approach a diameter of a half mile.

Subsidence craters and calderas open where the surface subsides as magma drains from beneath. Typically, they have vertical walls and sloping bottoms filled with rubble, or flat floors where lava has ponded. Repeated collapse produces compound subsidence craters. For example, Halemaʻumaʻu, the

Keanakākoʻi is a pit crater near the summit of Kīlauea. This photograph from July 19, 1974, shows lava pouring into Keanakākoʻi Crater and simultaneously erupting from a fissure at the base of the far wall.

traditional home of Pele, the Hawaiian fire goddess, is a large pit crater in Kīlauea caldera. Small subsidence craters are called pit craters; those larger than a mile or so across are called calderas. Most Hawaiian shield volcanoes have calderas at their summits.

During the main stage of shield growth, molten lava may erupt from a vent for months, or even years. It may pool in the vent, creating a lava lake. The chilled crust atop the molten lake continuously breaks and reforms in response to the convective circulation underneath. You can watch many of the phenomena of plate tectonics modeled in miniature in the restless crusts of lava lakes.

Kīlauea caldera is a large compound crater that opened as the summit of Kīlauea Volcano subsided.

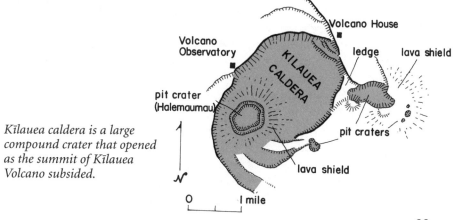

23

The crust of this lava lake at Kūpaianaha in Kīlauea's east rift zone tears into plates as molten lava churns beneath. —U.S. Geological Survey, Jim Griggs photo

Lava lakes also form where flows pour into pit craters. A lava lake several hundred feet deep may take decades to solidify, like the one that formed in Kīlauea Iki Crater on the Big Island in 1959. But the crust will be strong enough to support the weight of a person in a few days.

Lava Surfaces

Hawaiians long ago distinguished between two types of basalt lava flows: ʻAʻā lava has a craggy surface of jagged basalt blocks full of gas bubbles. It is almost impossible to walk across it without shredding your boots. Pāhoehoe lava has a billowy surface, ropy in places, looking almost smooth enough to be soft. Pāhoehoe surfaces on steep slopes typically assume the look of entrails. Pāhoehoe flows may also have large hummocks, called tumuli, that form where escaping gases heaved up plates of solidified lava. Fresh pāhoehoe lava has a paper-thin, shiny black glass crust that reflects light in iridescent colors.

Despite their differences in appearance, pāhoehoe and ʻaʻā lavas have the same chemical composition. In fact, both kinds of lava occasionally appear on different parts of the same flow.

Temperature is one of the more important factors that determines whether molten lava will develop an ʻaʻā or pāhoehoe. Hot basalt lava, very fluid, solidifies as pāhoehoe. Cooler lava, more pasty and stiff, tears itself apart through its own forward motion, creating rubbly ʻaʻā.

Pāhoehoe lava engulfed a school bus near Kalapana on Kīlauea in 1991.

Flowing lava does not heed stop signs. Between 1983 and 1985, when this photo was taken, ʻaʻā flows repeatedly entered Royal Gardens subdivision on Kīlauea. —U.S. Geological Survey, Jim Griggs photo

Dark ʻaʻā basalt and light pāhoehoe basalt drape a cliff and the flatlands below in Hawaii Volcanoes National Park.

The gas content of the magma also helps determine what kind of surface will develop. Gas makes the lava fluid, helping it to develop a pāhoehoe surface. Friction with the ground beneath causes shearing within the moving flow, especially if the lava is spilling rapidly down a steep slope. High shear situations with low gas content favor formation of ʻaʻā; low shear, with high gas content, favors formation of pāhoehoe.

Tongues of fluid pāhoehoe lava pour downhill in lobes no more than a few feet thick. As a hard crust forms on the lobe, the lava within flows out from under it, leaving a hollow tube. These develop one after the other, and overlap one another, so the flow of a single eruption may consist of a heap of thin, irregular layers, many of them hollow. Thin bands of reddened rubble generally separate them. Where separate pāhoehoe flows are stacked, they

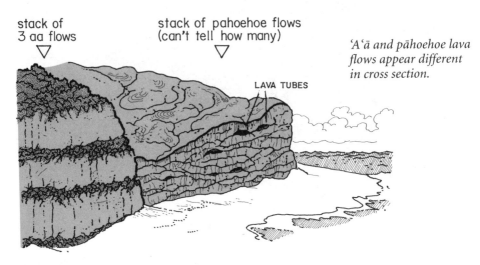

stack of
3 aa flows

stack of pahoehoe flows
(can't tell how many)

LAVA TUBES

ʻAʻā and pāhoehoe lava flows appear different in cross section.

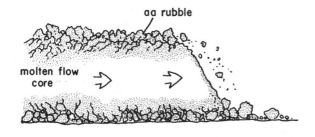

aa rubble

molten flow core

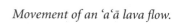

Movement of an 'a'ā lava flow.

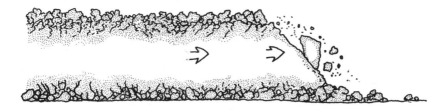

merge in such an intricately layered heap that it is difficult to tell them apart. A typical pāhoehoe flow is less than 10 feet thick.

In contrast, more viscous 'a'ā flows generally have a solid core, with thick red zones of rubbly basalt above and below. As the fluid interior of the flow advances, it carries rubble on the surface, rolling it over the front of the flow, then covering it. You can imagine it laying down its own tread as it crawls along. The individual flows are distinct. Typically, they are 10 to 30 feet thick.

Spectacular structures may form at volcanic vents. Many blobs of lava coughed out of a vent all at once are called spatter. Spatter piles up and forms ramparts, which stand like walls next to eruptive fissures. A towering lava fountain along a fissure may build a spatter cone several tens of feet high.

When lava pours into the ocean or a deep pool along a river, the water quenches the surface, forming a skin that swells up as molten rock accumulates underneath. The skin bursts, feeding a bulb of hot lava that in turn

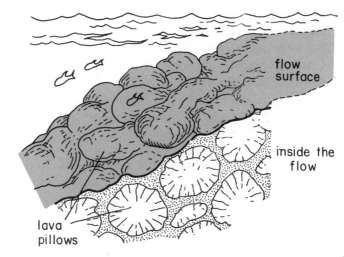

Pillow lava.

flow surface

inside the flow

lava pillows

27

Basalt columns.

grows its own skin, swells, and bursts. Over and over this process repeats, until the entire flow may appear to be a heap of rock bulbs. When you see them exposed in a cliff or roadcut, they look like a pile of pillows.

A single lava pillow usually is no more than a few feet in diameter. Chips of glassy lava and black sand shed from the growing pillow fill the spaces between it and its neighbors. Steam percolating through this debris may oxidize the iron in the basalt, staining it red.

In roadcuts, cliffs, and stream banks, you may see another feature typical of lava. The thick cores of ʻaʻā flows show neat palisades of vertical rock columns. Geologists have understood for more than a century that the columns express a pattern of shrinkage fractures in the rock. Some people who have watched lava flows in action report seeing those fractures form as the rock crystallizes, while it is still hot. Most of the shrinkage takes place in the transition from molten lava to solid rock. The further shrinkage that accompanies cooling is relatively minor, but it doubtless opens the fractures a bit wider.

Seen from above, the pattern of shrinkage cracks looks like the pattern of fractures you see in sun-cracked mud, or in crazed porcelain. The fractures outline polygons that have four to seven sides, most commonly five. Try counting the number of sides on basalt columns exposed in a roadcut or cliff.

Shrinkage cracks form perpendicular to the surfaces where the lava is losing heat. On most lava flows, those are the upper and lower surfaces, so the columns are vertical. If lava fills an old valley, the columns make a flaring pattern as they meet the valley walls at right angles. If magma fills a fissure to make a dike, the shrinkage cracks open at right angles to the walls, and the columns lie horizontally, like stacked firewood.

Lava Channels and Tubes

As the margins of a lava flow cool and harden, they confine the molten lava to a narrow channel. The centralized stream may become further confined as surges of liquid lava overtop the channel banks and chill into levees.

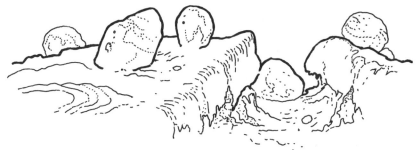

Lava balls along a channel in a pāhoehoe flow.

During periods of vigorous flow, big pieces may break off the walls of the lava levees and roll along in the stream of molten lava. The pieces grow larger as smooth coats of fluid lava chill around them, forming lava balls. You can see rounded lava balls scattered across the surfaces of many flows.

The channel at the center of a flow may also form a crust where the molten rock chills against the air. As molten lava beneath continues to flow and drains out from under the crust, it leaves behind a hollow in the flow—a lava tube. Some lava tubes are large enough that a person can walk through them, and some lava tubes are miles long. Most are very shallow, but some older tubes partly buried by younger flows extend deep into the ground.

The roof of a lava tube is an excellent insulator. Measurements taken at Kīlauea show the lava temperature dropped only a few degrees between the

A skylight in the roof of a lava tube reveals the interior spanned by a natural arch. Lavacicles hang from the ceiling. —U.S. Geological Survey photo

29

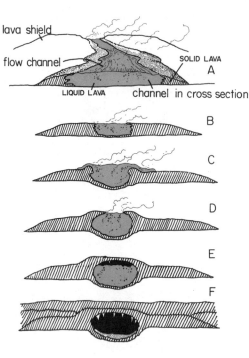

A common type of lava tube forms when the top surface of a lava flow cools and hardens. When the lava flow stops, a hollow tube remains.

point where it entered a roofed channel at Kūpaianaha, an active vent in the east rift zone, and the point where it emerged at Kalapana, 7 miles away. The insulating effect of lava tubes enables pāhoehoe lava to flow much farther than it would if it were exposed to cooling air. Without lava tubes, the shield volcanoes of Hawaiʻi would be steeper than they are, and would form much smaller islands.

When molten lava pours into a forest, the moist wood of the larger trees chills it around their trunks. If the flow level recedes, a pillar of hardened lava may remain standing; the trunk usually burns, leaving a hollow pillar. Geologists call such cylindrical monuments lava trees. If the flow level does not recede, a large tree trunk enclosed in lava may simply leave a vertical shaft called a tree mold. Hundreds of lava trees and tree molds adorn the surface of many pāhoehoe lava flows.

Age Dating

The vast bulk of the volcanic geology in Hawaiʻi began forming before people started recording history in the Islands. Geologists are like historians, but they usually cannot rely on eyewitness accounts to tell them when events happened. Instead, they depend on techniques that measure the natural decay of ancient radioactive materials. Perhaps the best-known technique involves carbon-14.

The chain of reasoning that leads to radiocarbon dates begins with the cosmic rays that constantly bombard the earth's atmosphere. They convert

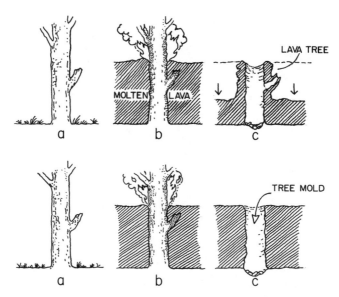

A lava tree or a tree mold will form after molten lava surrounds a tree.

some of the nitrogen in the upper atmosphere to carbon-14, which promptly reacts with oxygen to make carbon-14 dioxide. The cosmic bombardment maintains a constant level of carbon-14 dioxide in the atmosphere. Plants use it to build tissues, and animals eat plants, so every living thing contains carbon-14 dioxide at the same concentration as the atmosphere.

After living things die, they no longer absorb carbon-14 dioxide, and the amount in their tissues decays radioactively, turning back into nitrogen. The radioactive decay continues at a constant rate, so it is possible to determine how long ago an organism died by measuring how much carbon-14 is in the tissues.

Geologists studying volcanic rocks look for charcoal in volcanic ash or beneath a lava flow. If it appears that it was charred in the eruption, they can measure the carbon to obtain a radiocarbon date. This method works to an age of about 40,000 years. Several other more complex dating techniques work on older rocks, but not particularly well in the range between 40,000 and approximately 500,000 years. Older rocks are less trouble. The materials responding best to those methods are basalt or volcanic ash that is absolutely fresh, neither altered nor weathered.

Sinking Islands

As soon as they begin to grow, Hawaiian volcanoes begin to sink. The Pacific plate floats on the asthenosphere like ice on a lake. If you put a weight on ice, it depresses it into the water. Likewise, as a big volcano grows above the Hawaiian hot spot, it depresses the plate into the mantle. The volcano sinks until it is in floating equilibrium on the mantle, the same way a chunk

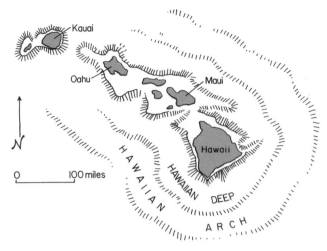

The oceanic crust around the Hawaiian Islands is bent into a moat called the Hawaiian Deep, with the Hawaiian Arch around the outer rim.

of wood sinks until the water can support its weight. Geologists call the flotation of the earth's lithosphere isostasy and refer to the slow subsidence of the Hawaiian volcanoes as isostatic sinking.

The Hawaiian Deep is a moat around the youngest Hawaiian islands, about 1,500 feet deeper than the surrounding ocean floor. It sank as the weight of the islands depressed the lithosphere. Both the ice and the lithosphere adjust to the added weight by flexing upward around the depression. The rise around the Hawaiian Deep is called the Hawaiian Arch.

After big volcanoes pass beyond the shield-building stage, they may lose as much as two-thirds of their elevation to isostatic sinking. The Big Island

Giant cliffs, or pali, mark the headwalls of slumps on the southern flank of Kīlauea. The weak southern flank of the volcano is sliding into the ocean.

is the youngest of the Hawaiian Islands, still growing and sinking faster than any of the others; the tide gauge at Hilo shows that the coast there is sinking about 1 to 2 inches every ten years. Oʻahu, an old island, long ago reached floating equilibrium with the mantle; tide gauges there show little sign of subsidence.

Even the islands that are in floating equilibrium are still sinking, although extremely slowly. They sink because the Pacific plate loses heat as it moves away from the east Pacific oceanic ridge. The ocean floor where they stand becomes denser as it cools, so it and the islands that stand on it float lower on the mantle.

Island Collapse

As Hawaiian volcanoes spread out and sink in response to the pull of gravity, huge land masses detach along rift zones and slide into the ocean. Some of them move slowly and enter the ocean in a thoroughly dignified, if ponderous, manner. Others plunge suddenly into the depths, with enough momentum to spread the slide dump more than 100 miles across the ocean floor. The largest of these landslides take place when a volcano grows to maximum size. They come around, on average, only once every 300,000 or 400,000 years.

Giant Hawaiian slides form because of the weak debris, thousands of feet thick, that makes up the submarine portions of all the islands. This is the

The northwestern flank of Kauaʻi, a shield volcano, collapsed into the ocean and created an enormous sea cliff that has eroded into the spectacular Nā Pali Coast. This view is from the Kalalau Trail, which precipitously hugs the coast for 10 miles.

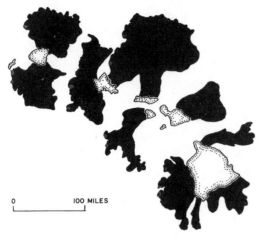

The dark areas are huge underwater landslides in the Hawaiian chain. Some of them have carried away a third of an island's landmass.

0   100 MILES

steam-shattered product of countless lava flows that have poured into the ocean. As an island grows, fresh solid lava covers up this sediment pile, weighing it down. Ultimately, the landmass grows so large that the weak material underneath gives way. A volcano swelling with magma may trigger landsliding by making the slopes of the island too steep.

Some landslides extend all the way to the ridge-crest rift zones and summits of volcanoes. Molten rock and fractures make these areas weak, creating natural break points for land separation. A third or more of an island can sink beneath the waves in a single slide.

Low sea cliff along the coast of Hawaii Volcanoes National Park. Nāulu sea arch is the result of wave erosion along fractures in the lava.

Giant landslides leave long cliffs where their headwalls cut the flanks of the volcanoes. Some of those cliffs are high on the slopes of the islands. Others rise thousands of feet out of the surf. All erode into spectacularly rugged terrain. Every main Hawaiian island has at least one coast where a towering cliff faces the ocean. Rocks exposed in many of those cliffs are full of vertical dikes, the filled fissures of defunct rift zones.

Small shoreline cliffs rising 10 to 50 feet above sea level are more common than the giant slide scarps. They are ordinary sea cliffs that develop as waves pull blocks of rock away from their bases. Then the undermined cliff collapses, dumping debris into the surf. Heavy seas sweep the rubble away, clearing the base of the sea cliff for more wave attack. As successive collapses drive the cliff landward, it leaves in its wake a gently sloping bedrock surface, the wave-cut bench.

Dramatic sea cliffs can form along volcanically active shorelines in a manner akin to the giant landslides mentioned earlier. As lava pours into the ocean, steam explosions and crashing waves pulverize the advancing flow, shedding vast amounts of black sand, spatter, and other fragments, which tumbles into the water. Eventually, the advancing lava overrides this debris, adding new land to the island. Periodic surges of lava cause the debris underneath to slide, collapsing the front of the flow. Geologists call this process bench collapse, and it is responsible for developing much of the low cliff on the southern coast of Kīlauea.

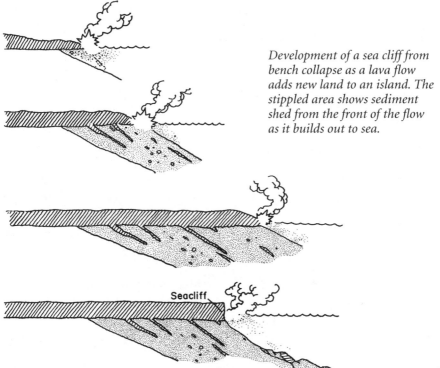

Development of a sea cliff from bench collapse as a lava flow adds new land to an island. The stippled area shows sediment shed from the front of the flow as it builds out to sea.

Seacliff

The top of this lava flow, on Lāna'i, is weathering into laterite soil and rounded residual stones. The roadcut is about 8 feet high.

Soils

In addition to isostasy and landsliding, ordinary weathering and erosion by flowing water help bring the islands back to the sea.

All rocks are fractured, most of them in regular patterns. Water seeps into the fractures and reacts with the rock, converting it into soil. Water attacks angular chunks of rock from two directions at edges, three directions at corners. So, an angular chunk of rock becomes rounded as the edges and corners weather more rapidly than the flat faces.

A good many Hawaiian soils enclose rounded residuals of the original rock; we call these residual stones. If the soil erodes, the residual stones lag behind to litter the surface. Many large rocks in Hawaiian streambeds became rounded not through stream transport but as residual stones.

⌐ Rocks weather into soil through a variety of physical and chemical processes that depend on climate. In Hawai'i, most of those processes work faster on the wet windward sides of the islands than on their dry leeward sides. Flows that erupted 200 years ago on the dry, leeward side of the Big Island still look almost perfectly fresh and support few plants. Flows that erupted only a few decades ago on the wet, windward side of the Big Island wear a coat of gray lichens and support a scattered growth of small shrubs and trees.

The chemical reactions between water and the silicate minerals in basalt contribute most to converting rock into soil. Basically, the silicate minerals turn into various types of clay, while most of their calcium, magnesium,

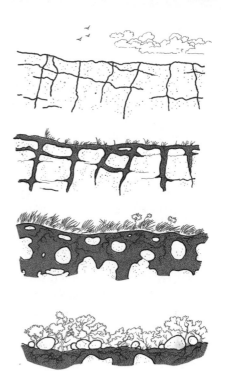

*Residual stones form as
lava weathers into soil.*

sodium, potassium, phosphate, and much of their silica go into solution. The silicate minerals swell as they turn into clay and no longer fit tightly together. The rock breaks apart the same way a wall would break down if the bricks in it were to swell, some more than others.

In wet regions, the rain washes the constituents out of the soil, leaving behind a residue consisting mainly of a clay called kaolinite, mixed in varying proportions with aluminum oxide and iron oxide, which stains it all red or yellow. The final result is a soil called laterite, the typical soil of the wet tropics.

Because laterite does not contain the soluble fertilizer nutrients calcium, magnesium, potassium, and phosphate, it is extremely infertile. It is difficult to fertilize laterite because the kaolinite clay does not retain fertilizer nutrients; they wash out readily, as the original nutrients did. Native jungle vegetation thrives on laterite because those plants are adapted to the infertility. They gain most of their nutrients from decaying litter on the forest floor.

The proportion of iron oxide, aluminum oxide, and kaolinite clay in laterite soil largely depends on the composition of the original bedrock. If the bedrock is rich in iron and poor in aluminum, the soil will consist mainly of iron oxide. If the bedrock is rich in aluminum and poor in iron, the laterite will consist mainly of aluminum oxide. Laterite that is extremely rich in aluminum and nearly without iron is bauxite, the only ore of aluminum. The Hawaiian trachytes consist mainly of feldspar minerals, which contain a lot of aluminum and no iron. Where trachytes are exposed on the wet sides of the islands, they weather into bauxite.

Some experts argue that mining Hawaiian bauxite could actually improve agricultural productivity by stripping off the most infertile top layers of the soil. They contend that the more fertile subsoil would support better crops. Many laterite soils are ten feet deep and more, so it is reasonable to argue that plenty would remain after the top is stripped off. But few people are ready to accept the idea that such mining would improve the appearance of the Hawaiian landscape or would qualify as the best use of its precious land.

Fluted Cliffs and Amphitheater Valleys

Besides the beautiful beaches and active volcanoes, one of the most striking aspects of the Hawaiian landscape is towering, fluted cliffs, including the famous Pali on Oʻahu. The wet northeastern sides of the Hawaiian islands all feature cliffs or pali, most of which originated as giant landslide scars. They are awesomely high and steep, their surfaces fluted with immense vertical grooves that look as if a monster had raked them with a set of giant claws. In fact, fluted cliffs covered with vegetation exist on most rugged tropical islands.

The fluted pattern reflects a unique aspect of tropical weathering and erosion. The warm water of the tropics dissolves rock more effectively than does rain in temperate climates. A comparison of soil and rock compositions shows that less than half the atoms in a typical sample of basalt remain after it is weathered into laterite soil. Many of the dissolved atoms enter the groundwater; some enter streams; all end up in the sea.

On a steep cliff face, naturally acidic rainwater clings to shady surfaces longer than it does to sunny, windy ones, where it readily evaporates. In time, the shady areas dissolve, leaving wide, deep pockets and gullies, while the sunny areas form ridges. The pattern smooths out into regular fluting, since the smaller ridges and gullies hidden in the shadow of the larger ridges remain moist and dissolve completely, leaving only the larger ridges and gullies behind.

In addition to fluting in cliffsides, spectacularly broad amphitheaters have formed at the heads of some large valleys, with long, thin waterfalls cascading down steep slopes. Other tropical settings have similar amphitheater-headed valleys, but not in more temperate climates. The Hawaiian valley walls are so steep that they can barely hold soil. The ridges between valleys are narrow, with knife-edge crests. Deep sediment deposits fill the valley floors.

The streams draining the Hawaiian islands are generally clear. They carry so little sediment that they seem unlikely to be entirely responsible for forming those huge valleys. They must have had some help. As with the fluted cliffs, rock dissolution must play an important role in forming the amphitheaters.

The unique amphitheater shape of most valleys in Hawai'i may be because most of the rainfall and moisture helping to dissolve them accumulates high on mountain slopes. When the upper end of a valley grows more rapidly than the lower part, it eventually will develop into an amphitheater.

Ancient volcanic calderas also play a role in the formation of some of Hawaii's most spectacular amphitheater valleys. Lavas altered and softened by hot gases while eruptions were still active make up the old calderas. Streams erode such soft rock easily, hollowing out the calderas into wide erosional bowls sometimes several thousand feet deep.

Frequent minor landsliding is another major erosional process on the wet sides of the Hawaiian islands. Deep soils developed on steep slopes of weak volcanic rocks are likely to slide, filling a valley floor. The extremely sharp ridge crests characteristic of the windward slopes are typical of tropical landscapes, where nearly continuous landsliding is an important process.

With all this steep topography, waterfalls abound. Even small streams are interrupted by delightful little falls and cascades. This is mainly because some rocks erode more easily than others. Volcanic ash erodes quickly, as do buried soils and the rubble zones above and below 'a'ā flows. Dikes and the massive cores of 'a'ā flows resist erosion and become the lips of waterfalls.

Running water uses abrasive particles of sediment to carve bedrock in the same way a sandblaster uses sand. Neither water nor wind alone can carve rock. The clear streams draining the wet sides of the islands are poorly equipped to carve solid masses of rock. When did you last see a postcard with a picture of a muddy waterfall?

Drowned Valleys and Tsunami

Streams carve valleys into the island as it sinks. Seawater floods the lower valley floor, and streams, landslides, and lava flows fill it, making a broad flatland framed between high canyon walls. You can mentally reconstruct the valleys' original depth by projecting the slopes of the valley walls downward to where they meet below the surface. In most cases, the sediment fill turns out to be hundreds of feet deep.

In due time, the sinking of the island carries so much of the valley below sea level that its upper reaches can no longer supply enough sediment to fill the floor. Maps of the underwater topography of the Hawaiian islands show many canyons that have completely drowned.

The sinking stream valleys of Hawai'i are potentially treacherous places to live because of the funneling effect they have on giant sea waves called tsunami. Most tsunami start when sudden movement on a fault shifts a large area of the ocean floor vertically, to the accompaniment of an earthquake. The movement of the seabed displaces a large volume of water and forms a wave that races across the ocean at speeds of several hundred miles per hour,

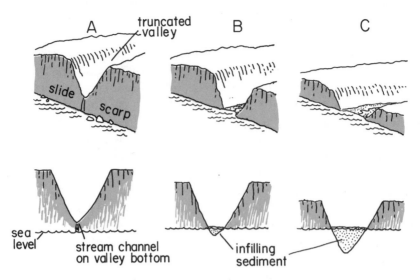

As a volcanic island sinks, the valleys fill with stream sediment. The result is a broad valley with a flat floor.

depending on the depth of the water. Such waves pose no problem to ships at sea, where crews do not notice their passing. But they slow down and build to monstrous heights when they approach land and enter shallow water, especially in bays and inlets.

Earthquake waves travel much faster than waves at sea, and seismograph stations monitor them constantly. Scientists can provide information on tsunami location and size hours before the water waves arrive. Earthquake waves travel about as far in a minute as a tsunami does in an hour. So a seismograph record of a distant earthquake on the ocean bottom can provide ample warning of a possible tsunami.

When seismograph stations issue a tsunami watch, Civil Defense personnel in Hawai'i prepare for action. In many cases, no tsunami develops, possibly because the fault movement shifted an area of the ocean floor horizontally. Only vertical displacements cause tsunamis. If a wave actually forms near the quake epicenter, the watch is upgraded to a warning, and sirens along the coasts in Hawai'i will sound, alerting people to evacuate. You can see these sirens, bright yellow horns or groups of dark green "doughnuts" on telephone poles, near many beaches.

A tsunami may arrive like a rapidly rising tide or like a rapidly falling tide. They have been compared to the sea flowing onshore like a broad river. In either case, very high and very low water levels occur at intervals of about 10 to 15 minutes. People have made the costly mistake of rushing out to catch the fish that lie exposed when a tsunami suddenly uncovers large expanses of the sea floor; the water level soon rises much higher than it was before it dropped. It is equally costly to stop running for high ground after surviving the first crest; the next several waves will probably rise even higher. After an hour or two, the tsunami ends with a series of progressively smaller waves.

Although earthquakes account for most tsunami that strike Hawai'i, the biggest waves arise from giant landslides suddenly dumping large chunks of a volcano into the ocean. Although none have been in historic time, geologic evidence gives some idea of their size: Geologists have discovered loose blocks of reef limestone 1,070 feet up the slopes of Lāna'i. These blocks were torn loose by a tsunami coming from the south about 100,000 to 105,000 years ago. Lāna'i stood higher then than it does today, so the wave may have risen even higher. The same tsunami probably stripped all the soil off the island of Kaho'olawe below a present elevation of 800 feet.

The effects of another monster tsunami in Hawaiian waters would be catastrophic beyond imagining, considering that the tsunami that wiped out downtown Hilo in 1946 was only about 20 feet high. It seems likely that most of the people who live in the coastal lowlands would drown—more than 95 percent of the population of the state. Such a wave would come with little or no warning because it would originate in local waters. Fortunately, monster tsunamis are so infrequent that there is no practical reason to worry about them.

Wind, Waves, and Beaches

Ordinary waves take their energy from the wind, then expend it in doing the work of shaping the coast. The prevailing northeasterly trade winds consistently drive the heaviest waves against the northeast coasts of the Hawaiian Islands, though storms far out at sea raise heavy swells that may come in from any direction.

Most people expect to see waves come straight in from the ocean, regardless of which way the wind may be blowing, or how the coast twists and turns. How do the approaching waves conform themselves to the outline of the shore?

Basically, they feel the bottom before they reach the land. When a wave enters shallow water, it slows down as it begins to drag on the bottom. Meanwhile, any part of the wave that may still be in deep water races ahead until it gets into shallow water, where it too begins to slow down. Imagine a wave approaching a coast obliquely: It will pivot like a line of marchers as one end slows in shallow water while the end still in deep water lunges forward.

Occasionally, waves approach the beach exactly head on, at a right angle. They then arrange the beach sand into a row of scallops with sharp points projecting into the surf, called beach cusps. The waves sort coarser sand into the cusps and leave finer sand in the low areas between them. Beach cusps last only as long as the wind is constant; when the wind shifts, the waves quickly erase them as they again wash onto the beach at a slight angle.

Anyone who has tried to snatch a choice seashell out of the waves quickly finds out what happens when the waves approach at a slight angle, as they

The beach at Wainiha, on the north shore of Kaua'i, continues across the mouth of the bay as a bar.

normally do. You try to grab the shell out of the incoming swash, only to miss it and see the backwash carry it out into the foaming water. When the next wave brings it back almost in reach, you have to walk a few feet down the beach to make the next grab. If you miss several times, you find you have moved a considerable distance down the beach. The wash of every incoming wave sweeps the shell obliquely onto the beach, and then the momentum of the moving water carries it farther down the beach in the backwash.

Every particle of sand on the beach is also moving down the coast. You can think of the beach as a river of sand flowing along the shore. On some days, the waves move the sand one way, on other days, the other way. Most beaches have a prevailing wave direction.

People commonly build walls, called groins, across the beach to trap the moving sand. Groins generally work very well. A row of groins converts a smooth, narrow beach into a much larger beach with a map outline something like the teeth on a ripsaw. However, the sand trapped on the growing beach never reaches its natural destination. If you look downshore from a set of groins, you will almost certainly find the beach there eroding, because the groins are starving it of the normal supply of sand.

Although breakwaters and piers do not look like formidable barriers, they trap sand as efficiently as groins, and also cause beach erosion farther downshore. Sand grains do not move under their own power—waves move them. Anything that interferes with the waves will affect sand movement.

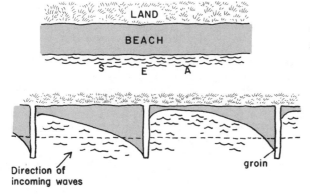

Groins can change the shape of a beach.

Direction of
incoming waves

groin

Anything that traps sand in one place will starve the beach somewhere farther down the line, causing beach erosion. Communities on some Hawaiian coasts import sand to maintain their beaches.

Whether a beach consists of sand, pebbles, or cobbles depends partly on the size of particles available and partly on the size of the waves moving them. In big storms, and on the windward sides of the islands, waves winnow out the small particles, leaving only cobbles and boulders on the beach. Sandy beaches are more abundant on the leeward sides, where waves are smaller. On all Hawaiian beaches, the smaller waves of calmer seasons carry sand onto the beach, burying the big rocks until large waves in the next heavy storm uncover them again. It is a long-standing pattern: Waves store the sand offshore during heavy weather, then spread it across the beach when the weather improves.

Lava from Kīlauea pours into the ocean along the south coast on the Big Island of Hawai'i. —U.S. Geological Survey photo

In 1991, lava from Kīlauea buried this famous black sand beach at Kaimū.

On the youngest Hawaiian shores the beaches are made of black sand. When molten lava enters the ocean, steam explosions blast the liquid rock into fine sand particles. Each grain is a jagged piece of volcanic glass. Waves and currents sweep the sand along the shore, where it collects in sheltered coves to make black sand beaches. The sand supply is not constant, so black sand beaches tend to wash away after a few centuries. Lava flows recently buried the most famous Hawaiian black sand beach, along the south shore of Kīlauea Volcano on the Big Island, while creating new ones nearby.

Where surf and currents erode fresh beds of volcanic ash, they separate grains of olivine from the lighter grains of other minerals. The olivine grains concentrate into green sand beaches that also contain black pyroxene.

On somewhat older Hawaiian shores, streams wash black basalt and red oxidized cinder to the coast. The basalt sand grains make beaches ranging from black to pale gray. The cinder may collect to form red sand beaches.

Coral reefs flourish along the oldest coasts, on islands that are no longer sinking very fast. Waves pound the reefs and break them into fine grains of beige to yellow calcareous reef sand, which collects on the shore to make the most stable, and most famous, beaches in Hawai'i.

Layers of cemented sand sloping up onto a beach from the waterline are beach rock. It differs from reef rock, which doesn't have such layers.

Seawater is slightly alkaline and does not dissolve calcite, the major mineral in reef sand. But beach sand above sea level is exposed to the rain, which is slightly acidic. Rainwater dissolves calcite from each grain of sand it wets. As

Beach rock on an ancient coral reef near Yokohama Beach, Kaʻena Point, Oʻahu. —Ronald K. Gebhardt photo

the solution of calcite and rainwater soaks deep into the sand, it precipitates calcite in the spaces between grains, cementing them together. The high beach becomes solid beach rock.

When sea level drops, wind and rain quickly erode exposed beaches. Beach rock is much more resistant than loose sand and may remain intact long after the surrounding sand has disappeared. Stony remnants of ancient beaches are common on all the older Hawaiian islands. Eroded beach rock shows many sloping layers. Each is a past beach face buried as the beach acquired more sand. The layers slope toward the ancient shore.

Coral Reefs and Sand Dunes

Beach rock typically overlies wave-cut benches in basalt, and in some places it overlies reef rock. The kinds of coral that build reefs live only in water consistently above 65 degrees. Many corals live in association with algae that require sunlight, so the water must also be clear and shallow. Other kinds of algae and many kinds of animals living on the reefs contribute to their development.

Corals are animals related to anemones and jellyfish. They live partly by snatching microscopic animals from the passing water and partly on the largesse of the algae that live in their tissues. The algae are photosynthetic plants that take in carbon dioxide, use the carbon in building their tissues, and release free oxygen. The coral use the oxygen for their own metabolism, and consume some of the algae. The algae, in turn, use the carbon dioxide the coral produces, and benefit from the shelter it provides.

Individual coral animals, or polyps, look like minute anemones, with tiny tentacles that wave in the water. Some of the tentacles are stingers. The polyps live in colonies where each little tentacled organism sprouts from a continuous membrane of coral flesh. They take dissolved calcium from seawater, combine it with some of the carbon dioxide they produce in their metabolism, and deposit it as calcite, the basic mineral matter of the reef. If you look closely at a piece of reef coral, you can see the little dimples in which the individual polyps nestled.

Corals spread locally by budding new polyps from the continuous membrane of tissue; they spread more widely by shedding enormous numbers of eggs into the passing seawater. The eggs develop into larvae, which drift in the water until the time comes for them to develop into polyps. Then they settle on whatever hard base they may encounter, where they attempt to start a new colony. The chances that any individual larva will survive all the hazards of the ocean are almost vanishingly small. This means that it takes a long time for reef corals to start on new volcanic islands. Even after 5 million years, Kaua'i has few very large reefs.

Established reefs face many hazards. Storm waves and tsunami may break them up or spread sand or mud across the coral, suffocating it. Agricultural chemicals washing into the ocean can kill coral, and so can sewage, which nourishes a smothering bloom of algae. Parrot fish use their heavy beaks to scrape calcareous algae from corals. Their incessant rasping produces large amounts of fine sand. Worms, sponges, starfish, sea urchins, and boring clams all bore holes in coral reefs.

The most devastating of the many hazards threatening coral reefs may be the changes in sea level that accompany the coming and going of ice ages. Sea level slowly drops as glaciers grow during an ice age, leaving the old reefs high and forcing the reef zone offshore to the new coast. Ice ages end suddenly, and sea level rises rapidly as the glaciers melt. The meltwater quickly submerges reefs beyond the depths at which corals and their associated algae can live. These fossil reefs, including some that grew during previous ice ages, girdle the submerged slopes of most Hawaiian islands.

Corals are highly competitive animals, so different species dominate in different localities in a reef. They come in shades of white, pink, yellow, brown, blue, purple, or black, but all fade to the white of bare calcite after the polyps die. Corals also come in many shapes, some massive, others rounded or branched like antlers. Sea fans look like flattened trees and grow in a rainbow of colors, and lettuce corals look like crinkled leaves. Mushroom corals are shaped like the cap of a broad mushroom several inches across, with many thin ridges radiating spokelike from a central stem. Corals growing in shallow

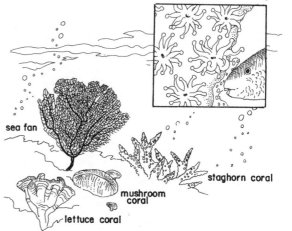

coral polyps close up, with parrotfish in background

Hawaiian coral reefs support abundant sea life.

sea fan

staghorn coral

mushroom coral

lettuce coral

water tend to be massive and rounded; those growing in deeper water, below the reach of waves, are more delicate and branching.

To the extent that the waves beat on coral reefs, they spare the main shorelines sheltered behind the reef. Without the barrier of the offshore reefs, waves would wash much of the beautiful sand in Hawai'i off the beaches.

Visitors to Hawai'i comment on the beautiful turquoise water. You see this color only in the shallow water beyond the coral reefs, and only when the sun shines. Two or three factors seem to have the most important effect in creating turquoise water: In the shallow water behind a coral reef, bright sunlight easily illuminates the white sand eroded from the reef. It reflects from billions of tiny particles of sediment and organisms suspended in the water,

Sea urchins boring into reef rock near Kīhei, Maui.

An ancient sand dune turned into solid rock and then eroded. Its interior shows characteristic crossbedding. Two miles northwest of Wailuku, Maui.

creating shades of turquoise. In deeper water, less light reflects off the bottom and the waves do not stir up sediment, so the water appears darker. Thus the different shades of color depend on the light, the depth of the water, and the amount of suspended sediment.

The sediment supply present in beaches and reef lagoons is truly enormous. It contributes to the formation of sand dunes as well. Some people associate sand dunes with deserts. In fact, sand dunes are just as closely associated with beaches in all climates, whether wet or dry.

Waves wash sand onto the upper beach at high tide or during heavy storms. When the upper beach dries in the sun, the sea breeze blows sand off it and into coastal dunes behind the beach. The dunes blow inland until they are beyond the reach of the ocean's strong salt spray, when plants can grow and stabilize them.

It is a marvel that sand dunes exist at all. Why does the wind sweep the sand into neat piles, instead of scattering it across the countryside? Imagine what would happen if the wind were blowing sand across an asphalt parking lot where you had laid a small blanket. The sand would catch on the blanket because it is soft; the grains bounce onto it, but not off. As a pile of sand accumulated on the blanket, it would catch more sand for the same reason that the blanket did, because it is soft. For a sand dune, softness is the essence of existence.

Changing sea levels complicate the life of sand dunes. When sea level drops, as in an ice age, coral reefs are exposed to the air and die. Wind sweeps across them, blowing large volumes of sand inland to build big dune fields. Then, when sea level rises again, the ocean floods the old reefs, greatly reducing the supply of sand. Plants then cover and stabilize the dunes.

Just as beach rock forms when infiltrating rainwater cements sand grains with calcite, calcareous sand dunes also turn into rock. Sand dunes typically are much thicker than beach deposits, so calcite cement generally penetrates only a little way into a dune, leaving the center soft. The outer rind of cemented sand blocks further penetration of rainwater.

Where erosion opens the interiors of old dunes, you may see many thin layers of hardened sand intersecting one another in an intricate pattern. Each layer is a former surface of a dune. Shifting wind causes dunes to slope one way, then another, accounting for the differently angled layering. Geologists call the overall pattern crossbedding.

As isostasy and plate drift continue, the volcanic bedrock of each aging Hawaiian island sinks into the sands of its own surrounding reefs and beaches, leaving behind a coral atoll—an echo of a vanished island. Ultimately the shifting sea floor carries the reef into such cool northern water that it dies and sinks. Each work of the hot spot, tens of millions of years old, disappears into the Pacific.

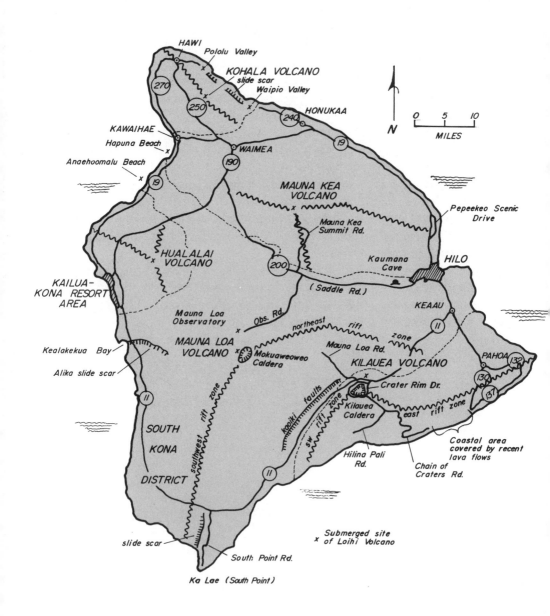

Geologic features on the island of Hawai'i.

2
Hawai‘i, The Big Island

Hawai‘i, the youngest and largest Hawaiian Island, is as large as all the others combined. It sprawls over an area the size of Connecticut, spanning 90 miles from north to south and 80 miles from east to west. Five large volcanoes coalesce to make the visible part of the Big Island; a sixth lies buried beneath the surface. The southern part of the island is still volcanically active and building out along much of the coastline. To the north, volcanism is in the waning stages. Of all the Hawaiian Islands, the Big Island shows the greatest diversity of rocks and landscapes.

Mauna Loa and Kīlauea

The largest volcano on Hawai‘i is Mauna Loa, sprawling over roughly half the island. Measured from its base on the sea floor, this mountain is one of the biggest on earth, with a maximum diameter of about 90 miles and a base-to-summit elevation of about 31,000 feet. It is exceeded only by a few older Hawaiian volcanoes, which have largely or completely subsided into the sea.

The summit of Mauna Loa, 13,677 feet, is at the northern rim of the volcano caldera, Moku‘āweoweo. Two main rift zones radiate from the caldera; the southwest rift zone meets the ocean west of Ka Lae, the southern tip of the island, and the northeast rift zone curves down the slope, ending in the rain forest about 15 miles south of Hilo. A third, more diffuse, rift zone radiates across the northwestern flank of Mauna Loa, with some fissures reaching as far as the foot of Mauna Kea, 20 miles north of Moku‘āweoweo.

Mauna Loa has erupted almost 40 times since 1832, but evidently is past its most active stage of shield building. The eruptions produce more lava in a shorter period of time than Kīlauea, its smaller neighbor to the south, which erupts more frequently. Mauna Loa appears to feed from a much larger magma chamber than the one supplying Kīlauea.

Mauna Loa is potentially the most threatening volcano in Hawai‘i. In the two centuries of historic record, individual eruptions on Mauna Loa have lasted as long as a year and a half, sending lava flows as far as 40 miles. Flows have threatened Hilo on seven occasions, most recently in 1984. If the great eruption of 1881 and 1882 were repeated, it would destroy hundreds of recently built homes and businesses in the city.

This towering cliff was formed when part of the western flank of Mauna Loa slid into the deep ocean near Ka Lae, the southernmost point of the island of Hawai'i.

Huge landslides have broken large pieces off Mauna Loa, leaving parts of the lower slopes very steep. Several systems of faults and large shoreline cliffs mark the boundaries of these enormous slides.

Mauna Loa's smaller southern neighbor, Kīlauea, is even more active, erupting more than 60 times since 1832. Because the surface of the volcano is young, it lacks a fully integrated stream drainage system. About 95 percent of Kīlauea's landscape is less than 1,500 years old.

Two rift zones extend from Kīlauea's caldera to the sea; the east rift zone, on the windward side of the volcano, slashes through dense rain forest, and the less active southwest rift zone stretches across a barren region called the Ka'ū Desert. No northern rift zone developed because the northern flank of Kīlauea is firmly buttressed against the larger and older mass of Mauna Loa. Kīlauea resembles a shoulder on the flank of Mauna Loa.

Kīlauea's eruptions threaten communities and roads time and again. Thousands of people live close to, or in, the lower part of the east rift zone, the area on the Big Island that has suffered the greatest property damage from eruptions. In 1960, cinder and lava buried the village of Kapoho near the eastern cape of the island. Pāhoehoe lava flows buried the community of Kalapana and the famous Black Sand Beach at Kaimū in 1990 and 1991. The potential for catastrophic loss increases as more people move into this accessible region.

Big slices of Kīlauea's south flank are slumping into the ocean along a series of faults, which creates a staircase topography on an enormous scale. Some of the scarps are as much as a thousand feet high. Occasional swarms of earthquakes confirm the continuing collapse. Submarine surveys show that the area measures 50 miles long and 40 miles wide. Some geologists think this terrain could someday slide catastrophically, dumping a large piece of the Big Island into the ocean and generating an enormous tsunami.

Earthquake studies suggest that Kīlauea is detached from the oceanic crust below and is slowly sliding south, with the southern flank of Mauna Loa. This adds credence to the idea that much of the southern part of Big Island, including Kīlauea, may one day slide or slump into the Pacific Ocean.

The Big Island's Dying Volcanoes

Hualālai, the Big Island's third historically active volcano, rises 8,300 feet on the central Kona Coast. It is older than Kīlauea and Mauna Loa and is in its late stage of activity. The extreme viscosity of the trachyte flows that were erupted during late-stage volcanism explains the unusually steep flanks of Hualālai. Viscous lavas tend to create steep slopes simply because they cannot flow down a gentle slope.

The trachyte lava flows also helped bury any caldera that may once have existed. Later, ash and flows of alkalic basalt covered most of the trachyte. About 25 percent of the alkalic basalt flows contain chunks of older rock, most commonly gabbro and dunite. Many late cinder cones and pit craters dot the volcano summit.

The volcano has two rift zones, which built it into a ridge about 20 miles long. The northwest rift zone is the longest and most recently active; it last erupted in 1800 and 1801. The southeast rift zone is much shorter, but part of it may be buried under younger flows from Mauna Loa. A few late-stage eruptions broke out on the northern flank.

Hualālai is a threat to thousands of people who live in the northwest rift zone and on the steep coastal slopes below, including the Kailua-Kona resort area. The volcano erupts at intervals of roughly once every 300 years. The last eruption was about 200 years ago. A swarm of earthquakes seemed to portend an eruption in 1929, but nothing happened.

Mauna Kea, at 13,796 feet, is the highest Hawaiian volcano. It dominates the landscape on the northern third of the Big Island. Like Hualālai, Mauna Kea has entered the late stage of its activity, in which cinder cones and flows of alkalic basalt have buried the original tholeiitic basalt shield. Geologists call the shield lavas of Mauna Kea the Hāmākua formation; the later alkalic rocks are the Laupāhoehoe basalts. The exposed Hāmākua lavas erupted during the dying stages of shield growth and include some alkalic basalts. The exposed rocks of the Hāmākua formation range from 150,000 to 100,000 years old.

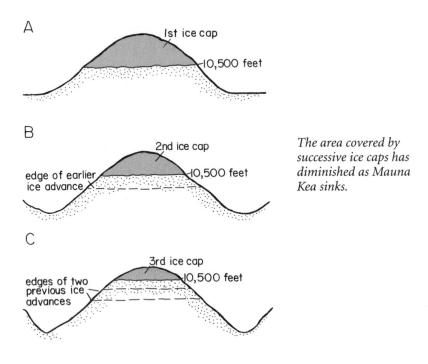

A

1st ice cap

10,500 feet

B

2nd ice cap

edge of earlier ice advance

10,500 feet

C

3rd ice cap

edges of two previous ice advances

10,500 feet

The area covered by successive ice caps has diminished as Mauna Kea sinks.

Laupāhoehoe eruptions began about 65,000 years ago. More than 300 cinder cones and lava flows of Laupāhoehoe basalt litter the summit region and flanks of Mauna Kea. Most of the cones are asymmetrical because the trade winds blew most of the erupting ash and cinder southwest. The latest eruptions, about 3,300 years ago, formed a complex of small cones a short distance south of the summit astronomical observatories. Those eruptions were recent enough to make further activity seem likely.

Scratched and grooved rocks, moraines, and other evidence of glacial activity record four general advances and retreats of an ice cap on Mauna Kea in the past 200,000 years. Because the volcano has been isostatically subsiding, evidence of the earliest ice advance extends to a lower elevation, 9,500 feet, than the residue of the latest advance, at 10,500 feet. In fact, all four ice caps probably reached to about the same original elevation around the sinking summit.

During its latest expansion, the ice cap covered almost 30 square miles, with an average thickness of 200 feet. The ice cap melted about 11,000 years ago. The ground beneath the summit is permanently frozen to a depth of 35 feet.

Laupāhoehoe volcanic rocks form a deep cover over most of the older Hāmākua formation rocks in the Mauna Kea shield above 7,000 feet. Lower down, many Hāmākua lavas are exposed or lie just beneath the surface. This lowland terrain, generally older and receiving more rainfall, has developed an abundance of stream gulches, especially on the windward northeast side of

Glacially scratched outcrops of basalt are common near the Mauna Kea summit.

the volcano, along the Hāmākua Coast. Deep gulches separate wide strips of the original volcano surface—fine land for cultivating sugarcane, grazing cattle, and building towns. Spectacular waterfalls make the Hāmākua Coast one of the most scenic drives on the Big Island.

Unlike Mauna Loa or Kīlauea, Hualālai and Mauna Kea show little evidence of large slides. These older volcanoes appear to be more stable, or perhaps a veneer of younger rocks covers the scars of past collapses.

Kohala, the Big Island's oldest volcano, forms a large peninsula extending north from Mauna Kea. Kohala is shaped like a ridge about 20 miles long. It grew along two prominent rift zones: one trending northwest, and a shorter one trending southeast. Lava flows from Mauna Kea cover part of the southeast rift zone. Kohala was at least a mile higher before it sank into the ocean. The present summit lies at 5,480 feet.

The older lavas on the main Kohala shield are the Pololū formation. The alkalic cinder cones and lava flows that veneer much of the volcano, especially along the summit ridge, are part of the Hāwī formation.

Lavas in the Pololū formation are the oldest on the Big Island. The internal magnetic fields, frozen into these rocks when they solidified, are aligned generally parallel to the earth's present magnetic field, so none could have erupted before 730,000 years ago, when the earth's magnetic field assumed its current orientation. The Pololū shield-building stage began to wane around 400,000 years ago. Basalt flows with 'a'ā surfaces became more common as activity tapered off.

Eruptions of alkalic Hāwī basalts began near the top of the volcano 250,000 years ago, with no significant lapse after the shield-building stage. Erosion had already cut gulches and canyons before Hāwī lava covered the lower slopes.

As far as anyone knows, Kohala last erupted about 100,000 years ago, soon after a set of parallel fault scarps developed along the rift zones, where they cross the summit. A strip of land dropped between two of the faults to make a depression called a graben.

A special set of conditions created one of the world's most unusual ecosystems at the summit of Kohala. During Hāwī volcanism, the trade winds blew huge volumes of ash downwind from active vents, spreading it across a wide area. The ash weathered into soil that contains a zone of impermeable clay, called a hardpan, just below the surface. Heavy rain falling on the nearly level ground at the summit could neither drain nor infiltrate, so it stayed on the surface. A swamp developed and supports thick beds of pale green sphagnum moss and such dwarfed native trees as the 'ōhi'a. Where the hardpan fractured, allowing soil water to seep down and emerge through nearby valley walls as springs, the infiltrating water eroded the ground below the surface, opening holes that later collapsed, producing large sinks.

Much of the northeastern flank of Kohala slid into the ocean sometime between 400,000 and 150,000 years ago, leaving a towering shoreline cliff more than 1,500 feet high. Gulches have since eroded into deep canyons with flat floors near sea level. Waipi'o and Pololū valleys are the largest. Other streams have yet to eat down through the face of the shoreline cliff to sea level: Their streams tumble as waterfalls across the cliff into the rough windward ocean below.

The largest coral reefs of the Big Island lie along the western shores of Kohala and Mauna Kea. Waves batter the coral into large quantities of white sand, which washes ashore to create such beautiful beaches as Hāpuna and 'Anaeho'omalu.

Ice ages left their mark on the highest mountain summits of Hawai'i, and also underwater. Dead reefs have been found as far as 4,000 feet below the waves on the submerged flanks of Kohala and Mauna Kea. Each reef corresponds to the end of a different ice age. They provide evidence of how much and how fast the volcanoes have sunk.

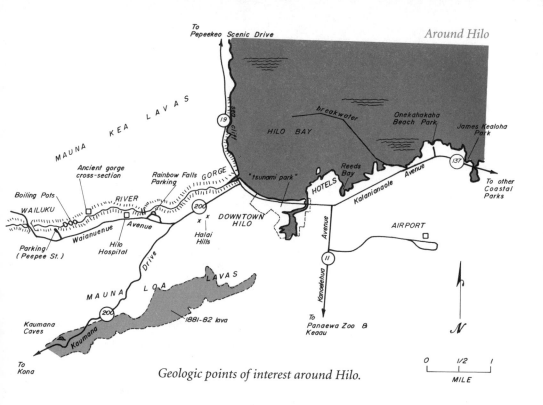

Geologic points of interest around Hilo.

Around Hilo

Hilo is nestled at the end of an ocean bay, where young lava flows from Mauna Loa meet older flows from Mauna Kea. The Wailuku River, crossing the northern part of the city, follows the approximate contact point of the two. Streams have dissected the older flows north of town, but not the younger flows south of town.

Part of Hilo stands on a lava flow that erupted from the northeast rift zone of Mauna Loa in the great eruption of 1881 and 1882. Most of the town stands on soils that weathered from lava flows and ash erupted between 24,000 and 14,000 years ago. A short distance south of the city, in the vicinity of the zoo and along Hawai'i 11 to Kea'au, the lava surface is only about 1,500 years old, barely weathered but already supporting a heavy plant cover.

The Wailuku Valley lies where lava flows from Mauna Loa lap against the older flank of Mauna Kea. The Wailuku River flows down the swale between the volcanoes. The river can make little headway in eroding the valley deeper because lava commonly flows into it from Mauna Loa, displacing the stream. The Hilo area contains striking evidence of the repeated filling and recutting of the Wailuku River gorge.

Rainbow Falls, along the Wailuku, is a short drive off Waiānuenue Avenue, near the Hilo Memorial Hospital. The plunge pool at the base of the falls undercuts the thick ledge of a basalt flow with an 'a'ā surface. If the light

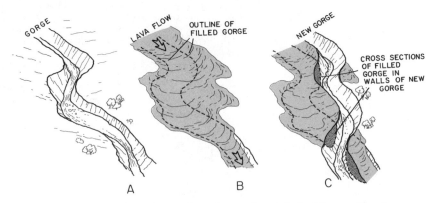

When lava from Mauna Loa flowed into the Wailuku River valley, the river cut a new valley, sometimes exposing sections of an older, earlier filled-in valley. The process is ongoing.

is right, you can see the curving base of the flow following the outline of an older Wailuku River bed that the lava filled. The flow erupted from the northeast rift zone of Mauna Loa about 10,500 years ago.

The massive cores of two older Mauna Kea ‘a‘ā flows, each with an excellent palisade of vertical columns, crop out in the walls of the gorge below the falls. At the bottom of the gorge, erosion has exposed a pāhoehoe flow with many lobes and small tubes.

Nearby Pe‘epe‘e Falls Street leads to the Boiling Pots. The rough outcrops in the river channel directly below the overlook are from a lava flow that erupted from Mauna Loa 3,500 years ago; it is the youngest lava in the river. Look carefully to see the lava pillows on the opposite bank. They were formed when lava filling the channel entered deep water.

Just downstream and a short walk from the Boiling Pots overlook, the Wailuku River passes through a narrow gorge rimmed almost completely with spectacular columns of basalt. The fractures outlining them are shrinkage cracks, which opened as the lava crystallized. The modern river is cutting a

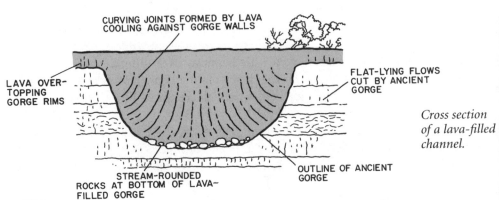

Cross section of a lava-filled channel.

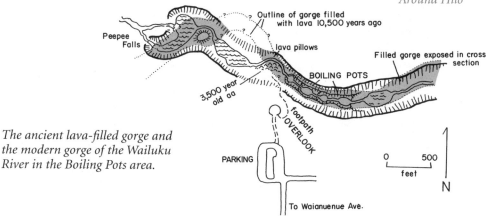

Outline of gorge filled
with lava 10,500 years ago

Peepee
Falls

lava pillows

Filled gorge exposed in cross
section

BOILING POTS

3,500 year
old aa

footpath

OVERLOOK

PARKING

0   500
feet

To Waianuenue Ave.

N

*The ancient lava-filled gorge and
the modern gorge of the Wailuku
River in the Boiling Pots area.*

channel almost exactly along the line of an earlier, filled gorge, exposing the columns over a wide area. The lava lining the walls of this cut is from the same flow that crops out at the lip of Rainbow Falls.

As the river cascades through the cut, it enters a chain of spectacular plunge pools, the Boiling Pots. In part, the river drains from pool to pool through underground fissures or lava tubes. Look for whirlpools along the banks where water funnels out of one pool and boils up into the adjacent pool downstream. These subterranean conduits take all the stream flow in times of drought.

If you venture downstream far enough to look past the lowest Boiling Pot, you can see part of the ancient lava-filled gorge exposed in a spectacular section along the left wall of the modern channel. The Wailuku River eroded the channel as we see it today directly along the plugged gorge in the Boiling Pots area, and across it farther downstream.

Boiling Pots.
—David Gianni
photo

59

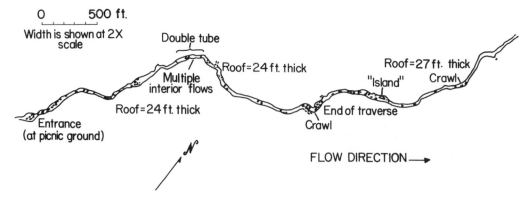

Kaūmana Cave lava tube. Adapted from Adams and von Seggern, 1969.

Kaūmana Cave

Kaūmana Cave, a large lava tube with easy access, is on Kaūmana Drive (the Saddle Road), about 5 miles west of downtown Hilo. The entrance is through a collapsed skylight in a small picnic area and county park. The lava tube was formed in the core of the lava flow that erupted in 1881 and 1882. Rubble blocks the upslope continuation, but people can walk or crawl downslope for nearly 3,000 feet.

The walls of Kaūmana Cave expose the internal structure of a pāhoehoe flow. In part, the layering in the upper walls was created by multiple spillovers of lava from the tube channel as it built up along the sides but before it was roofed over. Fracturing during cooling of the flow also created some layering.

The cave roof is 20 to 25 feet thick in most places. Most of the blocky rubble on the floor fell during or shortly after the eruption, when the skylight entrance also fell.

When the lava tube began, fast-moving molten rock filled the tube to the ceiling. Then, the level of the flowing lava dropped. During a pause in drainage, a steady stream of very fluid lava ran down the floor of the tube. From time to time, it slopped out to either side, building the smooth shelves you see near the cave entrance. Occasionally, roof blocks splashed down into the flowing lava, becoming partly embedded and coated. Later, the lava stream emptied, leaving the empty channel between the shelves.

Deeper inside the cave, the structure of the floor changes significantly. About 1,500 feet inside, a roof arches over the small channel, enclosing a small lava tube that grew in the same way as the larger tube around it. Farther along, the smooth pāhoehoe floor transforms into fine, granular ʻaʻā.

Many tiny stalactites and lava drip tracks adorn the upper walls and ceiling of Kaūmana Cave. They formed as lava dripped from the ceiling and dribbled down the walls as the level of the flowing lava, draining out along the floor below, dropped lower and lower.

60

The tree roots dangling from the ceiling are the base of a food chain that sustains a host of underground animals. Most of the larger creatures are insects and spiders. Although visitors have had a severe impact on these creatures of the cave, many animals continue to call this place home.

Not only have lava flows such as that containing Kaūmana Cave repeatedly entered Hilo, but volcanic vents formed in this neighborhood as well. An attractive neighborhood surrounds the Haili Hills in Hilo, south of Waiānuenue Avenue. These distant vents of Mauna Loa erupted 20,000 years ago. On the skyline north of town, pyroclastic cones date from the last gasp of activity along Mauna Kea's east rift zone.

From Hilo, you can look 25 miles northwest to the summit of Mauna Kea and see the white domes of the astronomical observatories. The volcano's gently sloping lower flanks show the profile of the older Hāmākua shield, which makes up most of Mauna Kea's total mass. The steep upper slopes, studded with Laupāhoehoe cinder cones and laced with younger flows illustrate the topographic impact of late-stage volcanism.

On clear days, you can see the broad and gently arching summit of Mauna Loa 30 to 35 miles to the southwest of Hilo. It appears low compared to Mauna Kea because it is farther away. You can also see some of the larger cinder cones along Mauna Loa's northeast rift zone. The most prominent is Kūlani cone, bristling with communications towers. Look for it in the rain forest on the lower flank of the shield.

Tsunami damage in downtown Hilo, 1946.
—National Oceanographic and Atmospheric Administration photo

If you look south from the top of one of Hilo's higher buildings, you can see pyroclastic cones and lava shields marking a trace of Kīlauea's eastern rift zone. It enters the sea at Cape Kumukahi at the eastern tip of the Big Island, about 20 miles southeast of Hilo.

Hilo Waterfront

Hilo Bay is a notorious tsunami trap. Most of them result from strong earthquakes around the Pacific rim, not local earthquakes. Waves build quickly as they climb the steep slope offshore from Hilo, and they lose little energy before reaching the shore. Once in Hilo Bay, they tend to reflect from one side to the other. Where reflecting wave crests meet, they combine to form a single, extraordinarily high crest.

The tsunami of April 1, 1946, smashed the port and killed ninety people. On May 22, 1960, waves killed sixty-one people. Waterfront property was devastated; the city simply cleared away the rubble and left what is today a wide, grassy strip referred to by local people as Tsunami Park. Only a few structures, including an exhibition hall, the Wailoa Center, and the county and state buildings, have been built there. Local phone books include maps of tsunami evacuation routes. Coastal hotels have contingency plans to evacuate guests to higher floors. Civil defense sirens are tested monthly.

In November 1975, a small tsunami caused by an earthquake along the island's south shore reached Hilo but caused little damage. Some more recent tsunami scares turned out to be false alarms, but they provided the opportunity for useful evacuation drills.

The Hilo shore is typical of young Hawaiian islands that have actively growing shorelines and little time for coral reefs to grow. The beaches are small and narrow, with many groundwater seeps.

A graceful black sand crescent follows the head of Hilo Bay between the old downtown district and Wailoa boat marina. The beach was more than twice as wide before shoreline construction and wetland filling reduced the sand supply.

Kalaniana'ole Avenue passes a string of beach parks. Rocks protect Reeds Bay from high surf. The bottom is rough in places, with patches of coral sand that washed in from Blonde Reef. Springs of fresh, cold water emerging from the bottom provide chilly but refreshing surprises when the weather is hot. One inlet nearby is called Ice Pond.

Keaukaha Beach Park is a section of rocky coast with sheltered water near shore. Freshwater springs feed Cold Water Pond.

Leleiwi Beach Park, sometimes called Richardson's Beach, is at the end of Kalaniana'ole Avenue. The shoreline is knobby pāhoehoe basalt with several sheltered inlets and a small black sand beach behind a breakwater. Freshwater springs cool the water, which typically is calm and full of marine life.

Pepeʻekeo Scenic Drive
15 miles round-trip from Hilo

Pepeʻekeo Scenic Drive begins 7.5 miles north of downtown Hilo. Follow Hawaiʻi 19 from Hilo toward Waimea, turn right onto Kulaʻimano Street, then follow the Scenic Drive signs along the shore back toward the city.

About 1.3 miles along the drive, the road crosses a stream gulch in a beautiful tropical forest. The stream tumbles through a natural arch, an eroded lava tube remnant, on the upstream side of the bridge.

Contrast the view of the young coastal lava flats on the eastern side of Hilo with the more deeply eroded landscape along the Scenic Drive. The youngest flows along the Scenic Drive are at least 100,000 years old. You can see how they have fared through centuries of weathering and erosion.

The Scenic Drive rounds Onomea Bay and passes Hawaiʻi Tropical Botanical Garden. Waves 35 feet high raced into Onomea Bay during the 1946 tsunami, wiping out a fishing village. The cliffs around the bay expose thin flows of ʻaʻā basalt, part of the Hāmākua formation. Their solid cores appear as pale layers separating the dark flow breccias. Look for the small red sand beach at the mouth of a stream that carries oxidized cinders into the bay.

The sea stacks at Onomea were left standing in the surf as waves eroded the coast. Their distance offshore shows how far the sea cliffs have retreated. Wave erosion took a notable step forward one day in 1956, when a natural arch at the northern end of the bay collapsed. Given time, they all collapse.

Hawaiʻi 11
Hilo – Hawaii Volcanoes National Park
30 miles

A visit to Hawaii Volcanoes National Park conveys a feel for the way all the older Hawaiian Islands were when their volcanoes were still active. You can think of it as a journey into the past of any Hawaiian Island. You may even see an eruption.

From Hilo, Hawaiʻi 11 heads south to Keaʻau, then climbs 4,000 feet in 20 miles to the summit of Kīlauea. Along most of the way, the road crosses lava flows, today heavily vegetated, that erupted from Mauna Loa's northeast rift zone only a few thousand years ago. Almost all the outcrops along the road expose black tholeiitic basalt containing large crystals of olivine.

You may see shallow holes about an inch across drilled into some of the outcrops and roadcuts. Those are places where core samples were taken to measure the magnetic field in the basalt, which is one way to estimate the age of a lava flow. The technique takes advantage of the slow movement of the

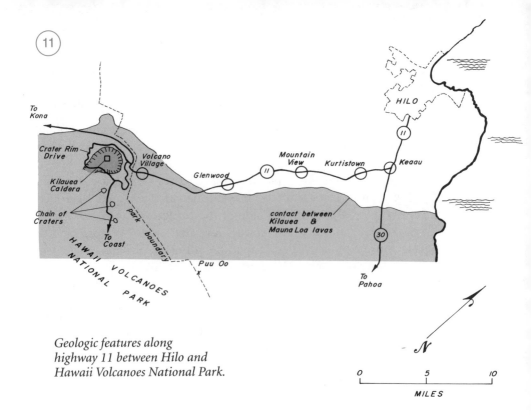

*Geologic features along
highway 11 between Hilo and
Hawaii Volcanoes National Park.*

earth's north magnetic pole. Knowing the history of this movement, geologists can determine when the flow erupted by finding the direction in which the rock became magnetized as it cooled.

Eighteen miles from Hilo, the highway passes Glenwood Park and Hirano Store, waysides in a long avenue of towering gum trees. A break in the woods provides a view of a gently sloping horizon to the south, Kīlauea's east rift zone. Pu'u 'Ō'ō, a prominent cinder cone about 700 feet high, dominates the horizon. It grew from a series of lava fountains that erupted over a period of three years, beginning in the early summer of 1983. Trade winds carried ash, cinders, pumice, and spatter downwind from the vent. The pit crater at the vent lies on the lower windward flank of the cone. Any steam you see is probably coming from this crater.

About 2.5 miles southwest of Hirano Store, the highway turns sharply where it passes the boundary between the well-vegetated lavas erupted from Mauna Loa and the lightly vegetated flows erupted from Kīlauea. The Mauna Loa lavas are several thousand years old; you can recognize them because they contain abundant, large crystals of green olivine. The Kīlauea flows are only 350 to 500 years old and contain few visible crystals of any kind.

From this point to the summit, the highway crosses only volcanic rocks erupted from Kīlauea. The vegetation near Volcano Village is lush not because the volcanic terrain is older but because it is covered by ash rather than

Eruptions in Kīlauea's east rift zone built Puʻu ʻŌʻō between June 1983 and July 1986. —U.S. Geological Survey photo

lava. Ash weathers more readily than lava, creating deep soil. This ash piled up during an unusually explosive eruption in 1790.

Where the road enters Hawaiʻi Volcanoes National Park, you can see Mauna Loa's northeastern flank, which is studded with tiny cones marking vents along the skyline. It is 15 miles away. You can also view Kīlauea caldera in its entirety from the porch of Volcano House, just inside the park.

A Short History of Kīlauea Caldera

Kīlauea caldera is about 2 miles wide and more than 3 miles long. At their highest point, the enclosing cliffs rise about 450 feet.

Halemaʻumaʻu, a low, broad lava shield capped with a pit crater, lies on the floor of the caldera. Lava overflowing from Halemaʻumaʻu and from nearby fissures has filled the caldera and begun to pour out at the southern end. It may bury the giant caldera completely in a few centuries. Then it will once again threaten the upper windward flank of the volcano, now protected by the walls of the caldera.

Most of the caldera formed shortly before, and perhaps during, the tremendous eruption of 1790. The eruption may have followed the sudden release of a huge volume of lava near the submerged base of the east rift zone. It drained the magma reservoir beneath the summit, causing the top of the volcano to erupt dense clouds of hot ash as it partly collapsed. Hawaiians who

Halemaʻumaʻu, the central vent of Kīlauea Volcano and traditional home of the Hawaiian fire goddess, Pele.

saw the eruption at close hand later described it in accounts suggesting it resembled the eruption of Vesuvius that buried Pompeii.

In 1790, the aliʻi (chiefs) Keōua and Kamehameha were at war and marching to battle near the southern tip of the island. Followers of Keōua, heading west from Hilo, reached the summit of Kīlauea on the established trail near nightfall. They camped close to the soothsayer's temple of Oalalauo on the northwestern rim, near the site of the Hawaiʻi Volcano Observatory. William T. Brigham, writing in 1909, described the eruption:

> During the night the eruption began by throwing out cinders and even heavy stones, the whole accompanied by the glare of molten lava, thunder and lightning. Fearstruck, the party in the morning did not dare go on, but spent the day in making offerings to Pele, but as on the next two nights there were similar disturbances they at last set out in three divisions. (Brigham 1909, 37–38)

Sheldon Dibble, writing in 1843, described what followed:

> The company in advance had not proceeded far, before the ground began to shake beneath their feet, and it became quite impossible to stand. Soon a dense cloud of darkness was seen to rise out of the crater, and almost at the same instant the thunder began to roar in the heavens and the lightning to flash. It continued to ascend and spread abroad until the whole region was enveloped, and the light of day was entirely excluded. The darkness was the

more terrific, being made visible by an awful glare from streams of red and blue light variously combined, that issued from the pit . . . and lit up by the intense flashing of lightning from above. Soon followed an immense volume of sand and cinders which were thrown in the high heavens and came down in a destructive shower for miles around. Some few persons of the forward company were burned to death by the sand and cinders, and others were seriously injured. All experienced a suffocating sensation upon the lungs and hastened on with all possible speed. . . .

The rear body, which was nearest the volcano at the time of the eruption, seemed to suffer the least injury, and after the earthquake and shower of sand had passed over, hastened forward to escape the dangers which threatened them, and rejoicing in mutual congratulations that they had been preserved in the midst of such imminent peril. But what was their surprise and consternation, when on coming up with their comrades of the center party, they discovered them all to have become corpses. Some were lying down, and others were sitting upright clasping with dying grasp, their wives and children, and joining noses [their form of expressing affection] as in the act of taking a final leave. So much like life they looked, that they at first supposed them merely at rest, and it was not until they had come up to them and handled them, that they could detect their mistake. . . . The only living being they found was a solitary hog. . . . In those perilous circumstances, the surviving party did not even stay to bewail their fate, but leaving their deceased companions as they found them, hurried on and overtook the company in advance at the place of their encampment. (Dibble 1843, as quoted in Brigham, 1909, 37–38)

Ash deposits on the dry western flank of the volcano preserve footprints left by the warriors and families.

In 1823, when Reverend William Ellis wrote the first detailed description of Kīlauea, the caldera was more than three times deeper than it is now. The center of the floor had sunk 300 feet. Ellis called the rim enclosing the collapsed area the Black Ledge. Spatter cones and lava lakes played continuously near the southwestern end of the central sink.

By 1829, lava had completely filled the inner sink and had flowed out across the Black Ledge. Then, in September 1832, the caldera collapsed again, forming a new black ledge. This event coincided with a small eruption in neighboring Kīlauea Iki Crater and on Uwēaloha (Byron Ledge), the land bridge separating Kīlauea Iki from the main caldera. The new sink slowly filled with new lava flows that erupted mainly from the southwestern side.

A tremendous earthquake jarred Kīlauea and the southern flank of Mauna Loa in April 1868. The caldera floor collapsed once again as magma drained rapidly into the southwest rift zone. William Hillebrand was one of the first to describe the event:

More than ⅔ of the floor of Kīlauea has caved in, and sunk from one hundred to three hundred feet below the level of the remaining floor. . . . The caving in of the floor seemed to be still in progression, for twice during

The lava mezzanine from the 1967–68 eruption is about halfway down the inner wall of Halemaʻumaʻu Crater.

our exploration of the crater, our nerves were disturbed by a prolonged heavy rumbling and rattling noise, as from a distant platoon-fire of musketry. (Quoted in Brigham, 1909, 106)

Within three years, overflowing lava lakes completely filled this new basin. The modern Halemaʻumaʻu lava shield had begun growing. The pit crater at the summit was much smaller than the one there now. By 1894, the lava shield had grown so high that the summit overlooked the southwestern rim of the caldera. Then, in July, the summit of the shield collapsed, converting Halemaʻumaʻu into a deep pit shaped like a funnel.

Activity resumed in November 1905, as lava again welled up into Halemaʻumaʻu. For the next nineteen years, Halemaʻumaʻu contained a continuously active lava lake 1,700 feet across in some places. It became world famous as the Hawaiian Fire Pit. Thomas A. Jaggar, of the Massachusetts Institute of Technology, established the Hawaiʻi Volcano Observatory on the rim of the caldera in 1912, to study this spectacular but gentle eruption.

The lava lake emptied suddenly in 1924, just as a huge volume of magma intruded the east rift zone. The walls of Halemaʻumaʻu Crater fractured and collapsed as molten rock drained out. Groundwater penetrated hot cracks in the crater walls and floor, boiled into steam, and erupted in a series of pow-

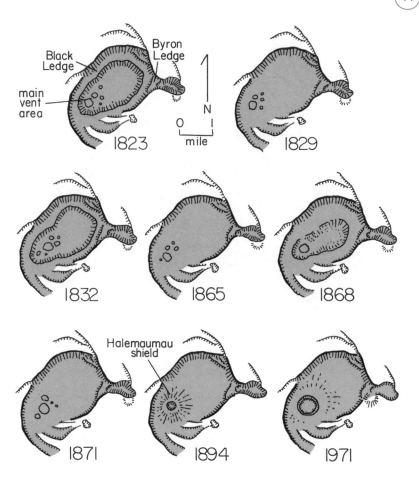

The changing shape of Kīlauea caldera since 1823.

erful explosions. Rhythmic steam blasts continued for a week and a half. When it was all over, Halemaʻumaʻu had deepened to 1,700 feet and widened to its present diameter. Blocks blasted out of the crater littered the southern caldera floor. They are still visible, except where younger lavas have covered them.

More than twenty eruptions have occurred in the caldera since 1924. Most of them lasted only a few days. One went on for eight months in 1967 and 1968, when the lava lake was again active in the Fire Pit. The lake had cooled and solidified when in September 1971 the central floor of Halemaʻumaʻu sank about 140 feet during an eruption in the southwest rift zone, which it apparently fed. The remnant of the lava lake of 1967 and 1968 left a narrow terrace around the walls of Halemaʻumaʻu.

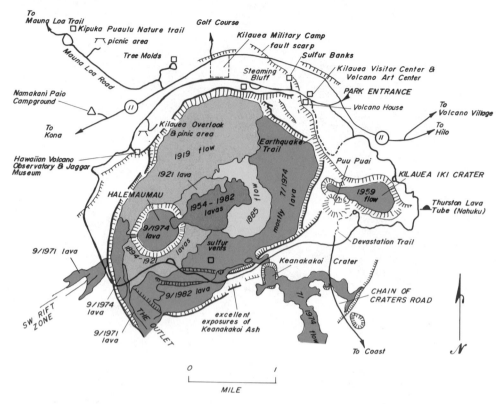

Geologic features along Crater Rim Drive in Hawaii Volcanoes National Park.

Crater Rim Drive Loop
11 miles

Crater Rim Drive is an 11-mile loop. You can drive the road in either direction. This description runs counterclockwise.

West of Kīlauea Visitor Center, the road descends one of the caldera's small, outer fault scarps and passes Sulfur Bank, an area of fumaroles and steam vents at the base of the scarp. Visitors a century ago used Sulfur Bank as a health spa, although no hot springs exist here because the ground is too porous and fractured to hold shallow hot water reservoirs.

Nearby Steaming Bluff is a grassy meadow laced with steaming ground cracks. Hot, acidic groundwater has killed all the deeply rooted plants. The road climbs another small fault scarp at the western end of Steaming Bluff, then passes Kīlauea Military Camp to reach the Hawai'i Volcano Observatory at Uwekahuna, a high rim of the caldera.

Southwest Rift Zone

The volcano observatory and Thomas A. Jaggar Museum of Volcanology afford spectacular views of Halema'uma'u and Mauna Loa. To the west, you

Sulfur Bank.

Steaming Bluff. The ridge on the horizon is Mauna Loa.

The Kaʻū Desert along Crater Rim Drive, showing gaping ground cracks in the southwest rift zone, which opened in 1868.

can see the upper part of Kīlauea's southwest rift zone as a series of huge fissures, fresh lava flows, and pyroclastic cones marching down the barren western flank of the shield. On a clear day, you can see Ka Lae, or South Point, 40 miles to the southwest.

Beyond the volcano observatory, the road winds across the southwest rift zone to the caldera floor. The change in plant cover between the wet windward slope and the dry rain shadow on the leeward slope is striking. But lower rainfall only partly accounts for the barren landscape of the Kaʻū Desert. It receives between 30 and 50 inches of rain a year, more than Seattle. Gases coming from fumaroles around Halemaʻumaʻu make the rain acidic, which helps explain the lack of vegetation.

You can see some of the huge fissures of the southwest rift zone from the road. They opened during an eruption that began hours after the major earthquake of 1868. Magma draining from beneath the caldera caused the ground to stretch and crack. The molten rock later erupted about 8 miles down the slope. As the magma drained away, Dr. Hillebrand watched the floor of the caldera sink.

Ash, cinder, and pumice from the 1790 eruption make up the strata in the walls of the fissures. Geologists call these deposits the Keanakākoʻi ash.

The Outlet is an opening in the southern wall of Kīlauea caldera. Lava flows have recently begun to pour through it.

On the Floor of the Caldera

Watch the roadcuts near the caldera floor for beds of Keanakākoʻi ash depressed beneath large blocks of basalt that fell on them. The pāhoehoe lava covering the flats where the road reaches the caldera floor was erupted from nearby fissures in 1971 and 1974. The road crosses a narrow strip of this lava. Closer to Halemaʻumaʻu, it crosses a much wider stretch of pāhoehoe lava that erupted in 1921 and is now weathered brown. Blocks of rock that were blasted out in the steam explosions of 1924 litter the older flow.

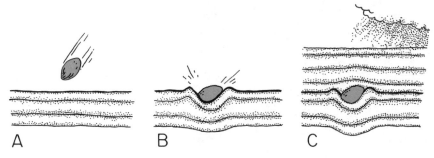

When volcanic bombs land on beds of soft ash, they dent it. Look for volcanic bombs in the Keanakākoʻi ash, with sagging ash layers beneath.

73

Blocks ejected during the steam explosions of May 1924 at Halemaʻumaʻu litter older lava flows. The caldera wall of Kīlauea, with Mauna Loa 20 miles beyond, is in the background.

Steam rises from ground cracks in 1919 lava along the trail near Halemaʻumaʻu Overlook.

Geologic map of lava flows and other volcanic deposits on the floor of Kīlauea caldera near the Halemaʻumaʻu Overlook. The dotted pattern shows the rubble ejected in the steam explosions of May 1924. The dotted lines show the spatter ramparts at the vents of various eruptions. Dashed bar symbols show the Halemaʻumaʻu and Byron Ledge trails.

The overlook at the rim of Halemaʻumaʻu Crater may be reached by a short walk from the parking area. It offers a good view of the prominent shelf of 1968 lava left on the crater wall when the floor sank in 1971. Look on the far wall for the dark fissure from the September 1974 eruption. A veneer of lava that erupted from this fissure covers most of the crater floor. Spatter cones on the bottom grew during a small eruption that followed an earthquake of magnitude 7.2 in November 1975. The small pad of lava on the crater floor to the right of the overlook erupted in April 1982 and is the youngest flow in Halemaʻumaʻu.

You can follow the trail another half mile beyond Halemaʻumaʻu Overlook to inspect spatter cones and ramparts along the rifts that produced the eruptions of 1954, 1975, and 1982. The spatter that was erupted in 1954 has already lost much of its glassy luster and has become yellow palagonite clay. The spatter of 1982 is still glassy and dark. Steam blowing through the vents during the waning stages of eruption oxidized the surrounding spatter to the color of brick. Small lava stalactites, lavacicles, dangle by the hundreds from the mounds of spatter overhanging the vents.

Look west from the spatter ramparts at the lower part of the cliff below and to the right of the observatory to see a broad strip of pale rock with two

An 8-foot-wide view of fragile 1982 pāhoehoe lava with a glassy surface lies next to Halemaʻumaʻu Trail north of Halemaʻumaʻu Overlook.

humps sandwiched between the dark lava flows. Some geologists believe this is a lumpy type of sill, called a laccolith, that was injected between the flows as a blister of magma.

Spatter ramparts line the fissures of the August 1971 and July 1974 eruptions near the beginning of the climb out of the central caldera on Crater Rim Drive. These ramparts are encrusted with yellow sulfur that crystallized from volcanic gases leaking from the vents. The vents mark the beginning of the east rift zone. Above the caldera floor, the road crosses a narrow strip of the flow that erupted in September 1982 then passes low fault scarps where you can view beautifully layered Keanakākoʻi ash.

Keanakākoʻi to Kīlauea Iki

Keanakākoʻi Crater is at one end of a chain of pit craters in the upper part of Kīlauea's east rift zone. Most, if not all the craters, are less than 1,000 years old. Many of them may have been formed during the great eruption of 1790.

The bottom of Keanakākoʻi Crater would be shaped like a funnel if the eruptions of 1877 and 1974 had not partially filled it. A short walk across the road from the crater leads to one of the rifts that erupted in 1974. Lava pouring from the small spatter rampart along the southern end of the rift flowed through a gully onto the caldera floor, swashing high against a channel wall as it rounded a bend.

In July 1974, lava poured down this wash near Keanakāko'i Crater, then drained onto the caldera floor, leaving lava on the channel walls.

Beyond Keanakāko'i Crater, the drive leaves the Ka'ū Desert and enters rain forest, passing roadcuts of deepening cindery pumice. These deposits accumulated in November 1959, when the trade winds blew shreds of lava from the top of a towering lava fountain erupting in nearby Kīlauea Iki Crater. The falling ejecta stripped leaves and small branches from the trees and partially or completely buried the forest for a mile downwind.

Much of the forest has recovered, but you can still see some of the eruption's effects along Devastation Trail, which begins near the intersection of Crater Rim Drive and Chain of Craters Road. The asphalt path is about a quarter mile long. It ends at Pu'u Pua'i Overlook, on the rim of Kīlauea Iki Crater.

The huge mound of cindery pumice dominating the skyline in the Devastated Area is Pu'u Pua'i, a cinder cone that grew on the downwind side of the 1959 vent. Hot gases and steam filtering through Pu'u Pua'i in the weeks after the eruption altered the composition of the cinder, giving the cone a golden yellow cap.

Beyond the intersection with Chain of Craters Road, Crater Rim Drive winds into a lush forest of tree ferns on the flank of 'Āi La'au, an enormous lava shield. The summit of 'Āi La'au collapsed 350 years ago and formed Kīlauea Iki. Near the top is Nāhuku, Thurston Lava Tube, smaller and less spectacular than Ka'ūmana Cave in Hilo, but safer and easier to visit. A well-lit, well-graded trail follows a segment of the tube for about 400 feet. It enters through the wall of a small pit crater and exits through a natural tube skylight. Other skylights downslope indicate that Nāhuku is at least a couple of miles long and may extend as far down the slope as the town of Mountain View. It is in one of the countless flows of pāhoehoe basalt that were erupted from ʻĀi Laʻau.

Puʻu Puaʻi as viewed from the Devastated Area. The lane between the shrubbery in the foreground was Crater Rim Drive before the 1959 eruption.

A nighttime lava fountain soars 600 feet above the vent as Puʻu Puaʻi grew during the 1959 eruption of Kīlauea Iki.
—National Park Service photo

The view from Kīlauea Iki Overlook on a clear morning is one of the finest in the park. The floor of Kīlauea Iki lies 400 feet below the overlook. The lava that flooded the crater in 1959 partially receded after the eruption, leaving a marginal subsidence terrace about 50 feet high—it looks like a gigantic, black bathtub ring. Uwēaloha, Byron Ledge, separates Kīlauea Iki from the younger central caldera, where Halemaʻumaʻu placidly fumes. The Jaggar Museum and Volcano Observatory perch on the cliffs overlooking Halemaʻumaʻu. Miles beyond, the gigantic shield of Mauna Loa dominates the horizon.

The 1959 eruption originated from the red vent at the foot of Puʻu Puaʻi, on the far wall of Kīlauea Iki. At times, the roaring lava fountain towered 1,900 feet high. The lava lake that pooled at its base is about 400 feet deep. Research teams have repeatedly drilled the thickening crust to study the lava's cooling and crystallization. By 1981, the crust was about 200 feet thick. The molten magma beneath had become so pasty that it simply graded into the crust.

A 4-mile loop trail leads to the floor of Kīlauea Iki. It passes the vent at Puʻu Puaʻi, the steaming drill holes in the crust of the lava lake, and the exposures of ʻaʻā basalt at the west end of the crater. The basalt is full of large crystals of green olivine, rare in lava erupted from the summit of Kīlauea.

The edge of the 1959 lava lake inside Kīlauea Iki became fractured when the central floor sank 50 feet as the lava cooled and drained into fractures and cavities.

To Kona

To Mauna Loa Trail

11

Crater

KILAUEA CALDERA

Kilauea Visitor Center

Rim

7/1974 flow

Drive

Thurston Lava Tube (Nahuku)

Lua Manu Crater

Puhimau Crater

To Hilo

Kokoolau Crater

5/1973 flow

Hiiaka Crater

To Hilina Pali

Koae

Faults

Puhimau Crater

Chain of Craters Road

Puu Huluhulu

N

Mauna Ulu

PARK

Kane Nui-O Hamo

Makaopuhi Crater

L A V A S

Napau Crater

Puu Oo

Muliwai a Pele (lava River)

Holei

U L U

Ke ala Komo Overlook

P U U   O O   F L O W S

BOUNDARY

M A U N A

0   3

MILES

Pali

Puuloa Petroglyphs

Paliuli

Chain of Craters Road is lined with major geologic features.

Sea Arches

Chain of Craters Road
20 miles one-way (dead end)

From Crater Rim Drive, Chain of Craters Road winds 20 miles southeast through Hawaii Volcanoes National Park to the barren, windswept south coast.

The first 5 miles follow the uppermost part of the east rift zone, passing the group of pit craters that give the road its name. In 1955, volcano specialist Gordon Macdonald watched Kīlauea's lower east rift zone eruption. In *Volcanoes in the Sea: The Geology of Hawai'i*, he wrote the only known eye-witness account of the formation of a pit crater:

At 4:03 P.M. on March 20, a sharp explosion in vent area *R* threw a billowing black cloud to a height of 500 feet, and several other similar but smaller explosions took place during the next hour. The closest ground observers were half a mile away, but a few minutes after the first explosion, an observer in an airplane found a new hole about 25 feet across in the ground surface at the site of the explosion. The interior of the hole was so brightly incandescent that it was difficult to see inside, but it appeared to be between 50 and 100 feet deep, with the walls overhanging so that it grew larger in diameter downward. The surface of the ground around the crater was covered with a thin layer of black, glassy ash and fine cinder. Undoubtedly, this black ash was the cause of the dark color of the explosion cloud. The volume of ash thrown out was only a very small fraction of the volume of the crater, and the ash was wholly glassy—no stony material derived from old rocks was present. Obviously, therefore, the old rocks formerly occupying the site of the crater had not simply been blown out by the explosion, but must instead have dropped in. (Macdonald and Abbot 1970: 88–91)

Within half a mile of the intersection with Crater Rim Drive, Chain of Craters Road crosses a narrow flow of pāhoehoe lava that erupted in 1974, then reaches Lua Manu, the first pit crater. The 1974 lava spilled into Lua Manu but did not fill it. Small spatter ramparts stand in the flow on either side of the road; they indicate that an eruptive fissure lies nearby, though it is not directly visible.

The next crater is Puhimau. More than 500 feet deep, its floor is covered with landslide rubble. Unlike little Lua Manu, it does not appear to have held molten lava since it collapsed. Koʻokoʻolau, the third pit crater along the road, also shows no sign of later eruptions. But the remnants of a large, mixed spatter and pumice cone on the rim indicate that it formed at an eruptive vent, unlike Puhimau and Lua Manu.

Hilina Pali Road

Near Koʻokoʻolau, Chain of Craters Road intersects Hilina Pali Road, a narrow lane that winds for 8 miles to a dead end. The first few miles pass fresh fault scarps in the Koaʻe fault zone. They have been active for at least 1,100 years. The road ascends the end of one fault, where the ground is warped rather than broken. The scarp nearby is so fresh that tilted trees grow from blocks that tumbled down its face.

Beyond the faults, Hilina Pali Road enters Kīpuka Nēnē, a patch of forest about 1,100 years old that stands like an island in a sea of much younger lava. With luck, you may see native Hawaiian geese, nēnē, at the picnic ground.

Seaward of Kīpuka Nēnē, the road winds across a pāhoehoe flow with abundant tumuli, passing through various grassy kīpuka. Any piece of land surrounded by younger lava flows is called a kīpuka. It may be only a few square feet or cover several square miles. The flow, which erupted sometime between 750 and 500 years ago, has weathered brown. It ends at the edge of Hilina Pali, a slump scarp 1,700 feet high. On clear days, you can see most of the south coast of the Big Island, including South Point. The coastal flats below are huge pieces of Kīlauea's southern flank that slumped toward the ocean.

Hiʻiaka and Pauahi Craters

Back near its intersection with Hilina Pali Road, Chain of Craters Road slices through a pair of small, prehistoric spatter cones, with the oxidized interiors well exposed. A small pullout where the road crosses a flow that erupted in May 1973 provides a good view of Hiʻiaka Crater, to the east. Part of the May 1973 flow was erupted from Hiʻiaka Crater, the rest from vents on either side of the road.

A path that starts at the Hiʻiaka pullout leads several hundred feet southwest across the 1973 flow. At the end of this path, a fault scarp forms a low cliff that cuts across a gentle slope, facing north. It is part of the Koaʻe fault zone, which cuts high across the seaward flank of Kīlauea to link the east and southwest rift zones. The Koaʻe fault zone separates the summit region of the volcano from the mobile southern flank.

Some geologists think the Koaʻe faults merge into a single horizontal fault at the base of the volcano. If so, the northeast rift zone of Mauna Loa could be the boundary between the detached part of the island, sliding on this fault, and the more stable area to the north.

A short distance past Hiʻiaka is Pauahi Crater, the largest of the easily accessible pit craters in the eastern rift zone. Three separate pit craters together form a composite depression 1,700 feet long and now about 350 feet deep. Three recent fissure eruptions partially filled Pauahi Crater with lava: two in 1973, the latest in 1979. Lava that erupted in 1979 covers most of the crater floor.

A deep lava lake formed during the 1973 eruption, then almost completely drained as activity waned, leaving the high bathtub ring of lava plastered against the western base of the crater wall. Most of the rubble on the ring fell because

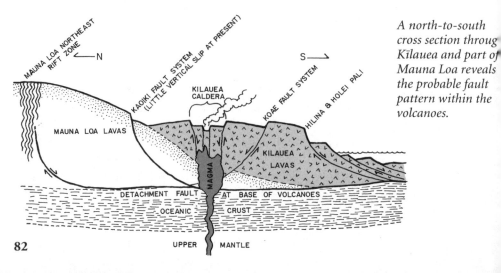

A north-to-south cross section through Kīlauea and part of Mauna Loa reveals the probable fault pattern within the volcanoes.

Pele's hair is the name given to brittle threads of golden glass stretched from falling lava droplets. Look for it in depressions and along embankments in the Mauna Ulu area.
—U.S. Geological Survey photo

of two earthquakes: a shock of magnitude 6.6 in 1983, and one of magnitude 7.2 in 1975.

The eruptive fissure of November 1973 cuts across the northeastern wall of Pauahi Crater. The fissure and spatter rampart near the rim, only a few tens of feet behind the visitor overlook, formed in May 1973. You can see lava trees, tree molds, spatter with iridescent glass and palagonite, Pele's hair, and other small features on the flow surface along the trial from the parking lot to the overlook. The fissure and spatter cones directly across the road from the parking area erupted in 1979.

Looking southeast from Pauahi Overlook, you will see two features on the horizon: Mauna Ulu, a lava shield that grew between 1969 and 1974, and Pu'u Huluhulu, a spatter and lava cone that grew about 500 years ago. Pu'u Huluhulu has been partly buried in the younger flows from Mauna Ulu.

Mauna Ulu and the Pu'u Huluhulu Trail

The short side road from the Mauna Ulu turnout follows the route of Chain of Craters Road as it was before the Mauna Ulu eruptions of 1969 to 1974. The trail to Pu'u Huluhulu starts from a parking area close to the point where lava covers the old road. Pu'u Huluhulu is about a mile away, Nāpau Crater another 6 miles.

The hike to Pu'u Huluhulu is rewarding, both scenically and geologically. You can see excellent lava trees, most of them in the flow that erupted from Pauahi in November 1973. The trail also provides close views of Mauna Ulu. Near the foot of Pu'u Huluhulu, the trail crosses pāhoehoe lava that erupted from Mauna Ulu. It is full of tree molds. The magnificent view north from

Lava trees are common in this November 1973 lava flow, along the trail to Puʻu Huluhulu.

the summit of Puʻu Huluhulu includes the summits of Mauna Kea, Mauna Loa, and Kīlauea.

Look in the saddle between Puʻu Huluhulu and Mauna Ulu for the pool of hardened lava that was fed from a flow erupting from the summit of the lava shield. Steep levees enclose this perched pond of lava. If you peer down the rift zone beyond Mauna Ulu, you can see a series of other lava shields and cones. The largest is Puʻu ʻŌʻō, 700 feet high. It grew during a series of eruptions between 1983 and 1986. In the middle distance, between Mauna Ulu and Puʻu ʻŌʻō, is Kāne Nui o Hamo, a lava shield that grew about 500 years ago. The western flank collapsed in the formation of Makaopuhi, the largest and deepest pit crater in the eastern rift zone.

The trail continues past the foot of Puʻu Huluhulu, over the pāhoehoe field around the northern flank of Mauna Ulu, and across the summit of a lesser lava shield on the eastern side. Then it leads on to the southern rim of Makaopuhi, skirts the eastern slope of Kāne Nui o Hamo through a dense ʻōhiʻa-fern forest, and ends at the rim of Nāpau Crater. The view of Puʻu ʻŌʻō from the end of the trail is excellent.

Hōlei Pali

Beyond Mauna Ulu, Chain of Craters Road crosses broad flows of pāhoehoe and ʻaʻā that rolled down the slope from the lava shield where it approaches the top of Hōlei Pali, the first great slump scarp on the south flank.

Powerful earthquakes in November 1975 caused seaward slumping of Kīlauea and Mauna Loa, resulting in this damage on Crater Rim Drive between Volcano House and Waldron Ledge.

Entrail pāhoehoe on Hōlei Pali.

At Muliwai o Pele pullout, the road crosses a beautifully developed flow channel in lava that erupted from Mauna Ulu in 1974. Surges of lava pouring through the channel cast many large lava balls onto the channel levees.

From Ke Ala Komo Picnic Shelter and Hālona Kahakai, you can take in an impressive view of the giant staircase topography of Kīlauea's collapsing southern flank. The cliffs below are slump scarps, the exposed part of the faults along which the broken pieces of the volcano slip. Except for those in the summit and rift zones, most of the volcano's earthquakes occur in this region. The magnitude 7.2 earthquake of 1975, centered near the coast, was the strongest in recent years. During that tremor, the shoreline dropped as much as 11 feet, and a big slice of the volcano slid 25 feet seaward. Even though part of the coast sank, the land area of Kīlauea grew.

Chain of Craters Road winds down the face of Hōlei Pali, a slump scarp about 1,500 feet high. Intertwined ʻaʻā and pāhoehoe lava flows that erupted from Mauna Ulu drape the face of the cliff. Like much lava that spills down steep slopes, this pāhoehoe has a peculiar form reminding imaginative geologists of entrails.

At the Alanui Kahiko pullout, you can see a short segment of the old Chain of Craters Road in a small patch of older ground surrounded by pāhoehoe lava that erupted in 1972. The road descends other small scarps and passes Puʻuloa, a large petroglyph site, before it reaches the coast. Wave erosion has carved a sea cliff, and in a few places has exploited weak zones of fracturing or rubbly rock to create sea arches.

Molten lava dribbling from the roof of a small lava tube formed these small lavacicles near the base of Hōlei Pali in the Mauna Ulu flow field.

Lava from an eruption in the east rift zone, 7 miles inland, has blocked Chain of Craters Road since 1986.

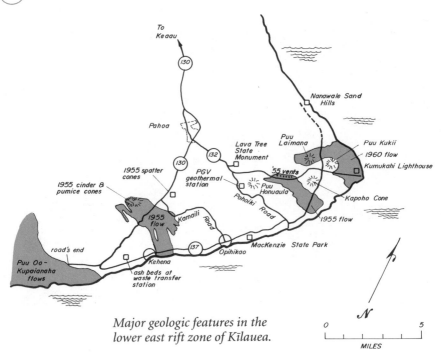

To
Keaau

130

Nanawale Sand
Hills

Pahoa

Lava Tree
State
Monument

Puu
Laimana

Puu Kukii

1960 flow

132

Kumukahi Lighthouse

130

1955 spatter
cones

PGV
geothermal
station

'55 vents

1955 cinder &
pumice cones

Puu
Honuaula

Pohoiki Road

Kapoho Cone

1955
flow

Kamaili

Road

1955 flow

road's end

Puu Oo-
Kupaianaha
flows

137

Opihikao

MacKenzie State Park

Kehena

ash beds at
waste transfer
station

N

0   5

MILES

Major geologic features in the
lower east rift zone of Kīlauea.

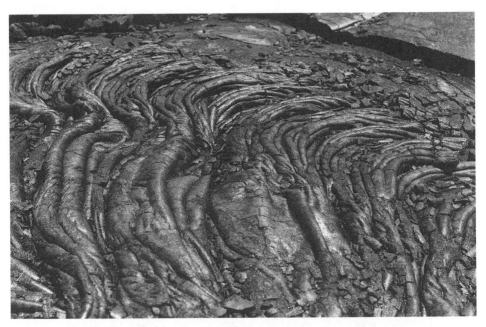

This fragile pāhoehoe lava erupted from Kīlauea's east rift zone in 1974 near the
coast south of Hilo; it flowed right to left. The field of view is about 8 feet across.

Hawai'i 130
Kea'au–South to the Coast
21 miles

Just south of Kea'au, Hawai'i 130 crosses weathered lava that erupted from the northeast rift zone of Mauna Loa within the past several thousand years. Enough soil has developed on it to support sugarcane.

About 1.5 miles south of Kea'au, the highway passes onto lava flows that were erupted from Kīlauea sometime between 500 and 350 years ago. The vents that fed these flows are about 20 miles to the west, near Kīlauea Iki. Very little soil has developed on this young terrain, so it supports few crops, only small trees and shrubs.

The plant cover becomes more dense and lush near Pāhoa. This does not mean that the rocks are older, merely that the climate is wetter because the crest of the east rift zone lies just south of town.

South of Pāhoa, Hawai'i 130 ascends the forested flank of the rift zone. You can see the first young volcanic features a short distance south of the rift crest, where steaming spatter cones appear east of the road. They grew during a large eruption that spread through the lower east rift zone in 1955, the first volcanic activity to strike Puna in 115 years.

Along the gradual descent from the ridge to the south coast, the highway crosses a huge flow of basalt that erupted in 1955. The rock has an 'a'ā surface and is full of glassy crystals of green olivine. It erupted from the pair of large spatter and cinder cones on the horizon west of the highway.

Pāhoehoe lava erupting from the east rift zone from 1986 until 1992 flowed all the way to the coast. It blocked Hawai'i 130, filled Kaimū Bay, and buried the town of Kalapana, a national park visitor center, and two black sand beaches. The Big Island gained more than half a square mile of new land. Hundreds of residents were displaced, but no one was hurt.

Hawai'i 137
Hawai'i 130–Kapoho
18 miles

The attractive red cinder bed of Hawai'i 137 branches off Hawai'i 130 near the edge of the young pāhoehoe flows that fill Kaimū Bay. This road hugs the shore most of the way to the island's eastern cape.

Between where it begins and the village of Kehena, the road crosses a flow of basalt lava with an 'a'ā surface that erupted around 250 years ago. The forest

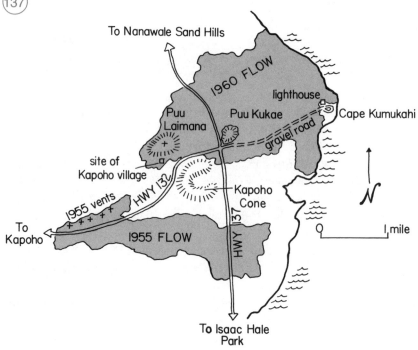

To Nanawale Sand Hills

1960 FLOW

lighthouse

Puu Laimana

Puu Kukae

Cape Kumukahi

gravel road

site of Kapoho village

Kapoho Cone

1955 vents

HWY 132

To Kapoho

1955 FLOW

HWY 137

0   1 mile

To Isaac Hale Park

Geologic features in the Cape Kumukahi–Kapoho area.

recovery is impressive, though not yet complete, despite the heavy rainfall in this area. Beds of volcanic ash that erupted in steam explosions are exposed in a cut along the truck siding at the Kalapana Solid Waste Disposal site. The ash beds have altered to the yellow clay mineral palagonite and the yellow iron oxide mineral limonite.

Just east of Kehena, Hawai'i 137 crosses rough lava that erupted in 1955. New houses are scattered across the flow. Farther east, the road crosses a much narrower tongue of 'a'ā lava that also erupted in 1955. It winds through the dense coastal forest west of 'Opihikao.

A prehistoric littoral cone, Pu'u Ka'akepa, rises seaward of the road less than a mile east of 'Opihikao. It grew at the end of a basalt flow that poured into the ocean. A quarry opened a good view of the layered interior, where steam partly altered the rocks.

MacKenzie State Park, with a thick forest of imported ironwood trees, is an excellent place to watch heavy surf pound resistant 'a'ā flows exposed in a shoreline cliff.

At Pohoiki Road, Hawai'i 137 turns toward Isaac Hale Beach Park, a landing for small boats, with a small beach of rounded stones and black sand. A spring of fresh, volcanically warmed water where swimmers and surfers like to rinse off is a short distance inland. The spring fills a large crack or perhaps a collapsed lava tube. Warm water also leaks into the bottom of the bay nearby.

Kapoho Ash Cone and Cape Kumukahi

Just north of Isaac Hale Beach, the road crosses a lava flow with an 'a'ā surface that erupted in 1955. North of the flow, Kapoho cone, a prominent ash cone about 350 feet high and well covered with plants, rises a short distance inland from the road.

Kapoho cone grew during a violent series of steam blasts in the east rift zone sometime between 400 and 1,000 years ago. The crater is shaped like a horseshoe because the trade winds blew the falling ash. A line of four small explosion pits cut across the floor of the crater.

The smaller ash and cinder cone that rises northeast of Kapoho cone is Pu'u Kuki'i (Pu'u Kūkae). It too is between 400 and 1,000 years old.

Hawai'i 137 reaches a four-way intersection between Kapoho cone and Pu'u Kuki'i. The gravel road to the right (east) leads to Cape Kumukahi, the easternmost point of land in the Hawaiian Islands. The cape is 1.5 miles away.

A large flow of 'a'ā basalt added new land to Cape Kumukahi in 1960. It nearly destroyed the Coast Guard lighthouse that stands at the end of the road.

Some of the small beaches along the rocky shore are green because they consist almost entirely of glassy crystals of olivine.

Nānāwale Sand Hills

North from the Kapoho cone–Pu'u Kuki'i intersection, the highway crosses the 1960 lava flow and continues unpaved into the Nānāwale forest. A bit more than 4 miles along, it passes onto a pāhoehoe flow that erupted in 1840 and today supports a thick forest.

The 1840 eruption broke out in several places. Activity began in the upper east rift zone, 20 miles to the west, too far away to disturb people in Puna. The action moved closer when new vents near Pāhoa erupted a flow that moved as fast as 5 miles an hour. It came on a Sunday, when people were gathered for church. The panic was terrific. The lava flow destroyed whole villages, and many people barely escaped. According to the Reverend Titus Coan, the intense night glow from molten rock and burning forest could be seen from 100 miles at sea. People 40 miles away could read by the light of the lava.

A pullout on the 1840 flow provides access to Nānāwale Sand Hills, two large littoral cones that grew where lava poured into the ocean for about three weeks. Wave erosion and sliding have consumed about half the mass of the cones, which originally stood 300 feet high.

Vents from the 1955 eruption along Hawai'i 132, west of Kapoho cone.

Hawai'i 132
Kapoho–Pāhoa
8 miles

Going west from the Kapoho cone–Pu'u Kuki'i intersection, you follow Hawai'i 132, which skirts the flank of Kapoho cone and passes Pu'u Laimana, a fresh cone made of ash, cinders, and spatter.

Pu'u Laimana grew in one month, during the 1960 eruption. You can hike the short distance from an unpaved pullout to the summit, where the view stretches along the east rift zone from Cape Kumukahi to the summit of Kīlauea. The tops of Mauna Loa and Mauna Kea are also visible.

Lava occasionally fountained to 1,700 feet when Pu'u Laimana was erupting in 1960. People could see it from Hilo, 12 miles to the northwest. The eruption buried Kapoho village, which used to be near the Pu'u Laimana highway pullout.

West of Pu'u Laimana, the highway crosses more lava from 1955 as it skirts a row of spatter cones next to the fissure that fed the eruption. Just west of this flow, Pu'u Honuaula appears a short distance south of the road; watch for the large cone covered with trees.

A body of hot, perhaps molten, rock lies at a shallow depth beneath Pu'u Honua'ula. Heat from this source powers the only geothermal generating plant in Hawai'i. The site is a short distance south of Hawai'i 132, on Pohoiki Road.

Lava Tree State Monument.

A lava tree,
about 6 feet high.

Tree mold along Mauna Loa Road, near the northern rim of Kīlauea caldera.

The Hawai'i Geothermal Project drilled its first test well, to a depth of 6,500 feet, in 1975 and 1976. The temperature of the rock at the bottom was 660 degrees. By 1982, a commercial generating plant provided electricity to more than two thousand homes; maximum production was about 3,000 kilowatts. Compare that to the 1,200,000 kilowatts that the geothermal plant at The Geysers in northern California generated for many years. Problems with noise and air pollution closed the well in 1989.

In 1993, Puna Geothermal Ventures opened two new wells at the base of Pu'u Honua'ula. This facility creates much less pollution than the first, and it generates 25,000 kilowatts, a substantial proportion of the Big Island's electricity.

Not far past the geothermal field, the highway enters Lava Tree State Monument, in a forest of towering albizea trees. These grow on a pāhoehoe lava flow that erupted in 1790 from a fissure where the restrooms now stand. The lava pillars formed when the flow first submerged the trunks of the trees, then subsided, leaving them cast in basalt. Much of the lava drained back down the fissure as the eruption ended.

Mauna Loa Road
Hawai'i 11–Mauna Loa Trail
12 miles

Mauna Loa Road is a scenic lane that begins on Hawai'i 11 in Hawaii Volcanoes National Park and winds along for 12 miles. It ends at the Mauna Loa Trail, a backpacking route that follows the northeast rift zone 18 miles to Moku'āweoweo caldera.

The tree molds off Mauna Loa Road are among the largest and deepest in Hawai'i. They preserve the shapes of mature acacia koa trunks encased in pāhoehoe lava that erupted from Kīlauea hundreds of years ago.

About 1.5 miles from the tree molds, Mauna Loa Road reaches the base of a long slope where Kīlauea and Mauna Loa meet. Kīpuka Puaulu is a patch of mature native forest growing on a deposit of volcanic ash that fell 2,200 years ago. Flows that erupted between 300 and 400 years ago surround it. A beautiful nature trail loops for a mile through the kīpuka, allowing visitors to explore a Hawaiian native upland forest, an endangered ecosystem.

Geologists adopted the Hawaiian word kīpuka to refer to areas of older land surrounded by younger lava flows. Volcanoes as active as Kīlauea and Mauna Loa have countless kīpuka. They provide isolated island habitats for many kinds of plants and animals, which helps explain why the Big Island supports so many species that live nowhere else.

Kīpuka Kī is another stand of old trees isolated by younger flows. The road continues through young koa woodland and scrub to an elevation of 6,700 feet on the southern flank of Mauna Loa.

The end of the road provides a spectacular panorama of Kīlauea's summit area. A short walk along the Mauna Loa Trail leads across sparsely vegetated pāhoehoe flows that erupted from the northeast rift zone, which is only 4 miles away as the crow flies, or about 8 miles across rough lava along the trail.

Hawai'i 11, Māmalahoa Highway
Kīlauea–Kailua-Kona
98 miles

From the summit of Kīlauea, Māmalahoa Highway rounds the southern tip of the Big Island where Mauna Loa's southwest rift zone enters the sea, and winds up the steep Kona Coast to the Kailua-Kona resort area.

Hawai'i 11 crosses the Sulfur Bank fault scarp, the northeastern edge of Kīlauea caldera, near the intersection with Crater Rim Drive in Hawaii Volcanoes National Park. Steaming Bluff and the crater area suddenly appear in the panorama south of the highway. The highway curves past Kīlauea Military Camp, a recreation area for federal personnel, and the road to Volcano Golf Course. Another caldera fault scarp rises as a low wall north of the highway. The top of this scarp is the highest point on Kīlauea, a bit more than 4,000 feet above sea level.

Ka'ū Desert and Mauna Iki

Hawai'i 11 descends the western flank of Kīlauea for 20 miles after the junction with Mauna Loa Road. Watch about 7 miles south of Nāmakani Paio

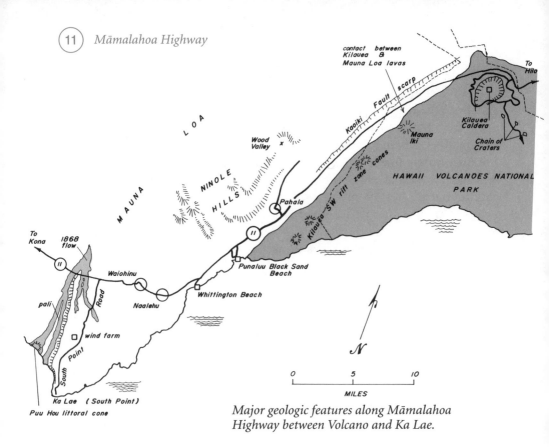

Major geologic features along Māmalahoa Highway between Volcano and Ka Lae.

Campground for the Kaʻū Desert pullout, south of the highway. Several lava balls litter the ʻaʻā flow near the pullout.

A trail leads from the Kaʻū Desert pullout to Hilina Pali Road. The first half mile or so crosses a basalt flow with an ʻaʻā surface that erupted from the northeast rift zone of Mauna Loa 280 years ago. Beyond, it crosses a flow of pāhoehoe basalt that erupted from Kīlauea 420 years ago. Ash and grit from the fearful eruption of 1790 cover most of the flow. The wind occasionally uncovers fossilized footprints of Hawaiians caught in that eruption—please do not disturb any you might discover. Look in the 1790 deposit for the rounded pellets of ash that geologists call pisolites. They may have formed as electrostatically charged ash clumped together inside the eruption clouds.

A walk of 2 miles from the parking area takes you to the summit of Mauna Iki, a lava shield that grew in eight months during 1919 and 1920. Some of the largest examples of tumuli in Hawaiian lava are near the this trail, at the base of Mauna Iki.

The summit of Mauna Iki provides a diverse panorama of cones, fissures, and lava flows along Kīlauea's southwest rift zone. A small pit crater at the top of the lava shield opened when a lava pond drained away beneath it.

One of the Twin Pits along the Mauna Iki Trail, about 6 miles from the Ka'ū Desert pullout.

The continuation of the trail past Mauna Iki, all the way to Hilina Pali Road, crosses a barren landscape that features pit craters, vent cones, embankments of Pele's hair, sand dunes, and fresh fault scarps.

The steep slope less than a mile north of the Ka'ū Desert pullout is a fault scarp in the Ka'ōiki fault zone, where an earthquake of magnitude 6.6 struck in 1983. The origin of the Ka'ōiki faults is unknown. Perhaps they began as landslides in the flank of Mauna Loa before Kīlauea grew large enough to provide support. But that would not explain why recent movement in the Ka'ōiki fault system has been horizontal, not vertical. Perhaps differential swelling of the magma chambers beneath Kīlauea and Mauna Loa creates the stress that moves the faults.

Nīnole Hills

Watch south of the road in the area west of the Ka'ū Desert for occasional glimpses of cinder cones in the southwest rift zone of Kīlauea. Near Pāhala, to the north, the Ka'ōiki faults die out, and the flank of Mauna Loa becomes a rugged cluster of steep hills and valleys. These are the Nīnole Hills.

The rugged landscape of the Nīnole Hills shows that they are older than the less eroded landscape around them. Most of Mauna Loa's surface formed in the last 4,000 years, but this part of the southern flank is 100,000 to 200,000 years old.

Formation of the Nīnole Hills. Color shows landslide scarps in the Nīnole Hills.

Mauna Loa before 200,000 years ago.

Collapse of Mauna Loa to the north.

Rebuilding of north flank, collapse of south flank.

Mauna Loa and Kīlauea grow to their present sizes.

The lava flows in the Nīnole Hills are shield basalts like the rest of Mauna Loa. Why, then, do the hills look so different from the rest of the volcano? Many geologists believe the Nīnole Hills are eroded remnants of an older version of Mauna Loa. The mountain to the north collapsed thousands of years ago, leaving the Nīnole Hills behind as a tall ridge. Since then, the young shield of Mauna Loa grew back to tower over the landscape. Flows from the modern volcano have not yet covered these patches of the older shield.

The steep, seaward-facing cliffs of the Nīnole Hills also show that a giant landslide must have torn away their southern flanks. The northeast coast of Kohala, unquestionably an eroding slide scarp, has a similar cliff and canyon topography. The difference is that Kīlauea has grown to cover the slide debris downslope from the Nīnole Hills, while the debris from the north flank of Kohala remains largely uncovered, making the situation there much easier to interpret.

Warm, moist air from the ocean rises high against the flank of Mauna Loa to maintain a nearly constant cloud cover above Pāhala and the Nīnole Hills. It explains the Hawaiian name for Mauna Loa's southern flank—'Āinapō, the land of darkness.

Thick soils weathering from ash beds tens of thousands of years old and sediment washing down from the Nīnole Hills support the sugarcane around Pāhala.

During the great earthquake of 1868, which was centered nearby, a huge debris avalanche burst from the walls of Wood Valley, a few miles upslope from Pāhala. It swept away ten houses, and killed thirty-one people and hundreds of cattle. Today, plant growth covers both the avalanche and the scar. The Reverend Titus Coan visited the avalanche shortly after the disaster and wrote a description of it. William T. Brigham quoted his report in *The Volcanoes of Kīlauea and Mauna Loa on the Island of Hawai'i*:

> I went entirely around it, and crossed it at its head and center, measuring its length and breadth which I found were severally three miles long and a half mile wide. The breadth at the head is about a mile, and the ground on the side-hill, where the cleavage took place, is now a bold precipice sixty feet high. Below this line of fracture the substrata of the earth, consisting of soil, rocks, lavas, boulders, trees, roots, ferns, and all tropical jungle, and water, slid and rolled down an incline of 20° until the immense masses came to the brow of a precipice near a thousand feet high, and here all plunged down an incline of 40° to 70° to the cultivated and inhabited plains below. The momentum acquired by this terrific slide was so great that the mass was forced over the plain, and even up an angle of 1° to 30°, at a rate of more than a mile a minute. In its course it swept along enormous trees, and rocks from the size of a pebble to those weighing many tons. Immense blocks of lava, some fresh as of yesterday, and others in all stages of decomposition, were uncovered by the slide. The depth of the deposit on the grass plains may average six feet: in depressions at the foot of the precipice it may be thirty or even forty feet. (Brigham 1909, 114)

Punaluʻu and Whittington Beaches

About 5 miles south of Pāhala, Hawaiʻi 11 passes the Nīnole turnoff to Punaluʻu Beach Park. This is an attractive beach, fringed with palm trees in the finest Hawaiian tradition. The beach is black sand—particles of glass shed from ancient lava flows that entered the sea nearby. The beach was larger before tsunami pounded this shore in 1868, 1960, and 1975, removing much of the sand.

Almost 10 miles south of Pāhala, Hawaiʻi 11 crosses flows of ʻaʻā lava that erupted from Mauna Loa's southwest rift zone and passes Whittington Beach County Park.

Plantation workers in days gone by loaded raw sugar onto barges at a landing at Whittington Beach. The landing closed around 1940, when improved trucks and highways provided a better method of transportation. The tsunami of 1946 destroyed the port facilities, as did the tsunami of 1975. The shore here is rocky, and fish abound offshore. You may catch a glimpse of a sea turtle.

The top of the grade overlooking Whittington Beach provides an impressive view east toward Kīlauea, which looks like a huge shoulder in the flank of Mauna Loa. You can easily see the series of scarps in the slumping seaward flank.

West of Whittington Beach, the road crosses the flank of Mauna Loa's southwest rift zone. You will not see any sign of recent volcanic eruptions. Paliomāmalu, a huge slide scarp a few miles west of Waiʻōhinu, diverts lava flows, and the steep western flank of Mauna Loa also provides some protection for local residents by drawing lava away. Despite substantial natural protection, flows of ʻaʻā basalt did approach in 1916 and 1926.

Waiʻōhinu and the Great Earthquake of 1868

The most severe shaking reported in the great earthquake of 1868 came from people in Waiʻōhinu, no doubt because they were closest to the epicenter. No seismographic instruments existed then, so we cannot determine the exact magnitude of the earthquake. A recent estimate places it around 8 on the Richter scale—enormous. The quake came a day after a small eruption at Mauna Loa's summit. Magma moving inside the volcano may have played a role in triggering the disaster.

William T. Brigham gave a chilling eyewitness account of the eruption and earthquake, which began with a series of foreshocks the day after the eruption and culminated with the main shock at 3:40 P.M. on April 2:

> March 27, 1868, about half-past five in the morning, persons on the whale ships at anchor in the harbor of Kawaihae saw a dense cloud of smoke rise on the top of Mauna Loa, in one massive pillar, to the height of several miles, lighted up brilliantly by the glare from the crater Mokuaweoweo. In a few hours the smoke dispersed, and at night no light was visible. . . .

The shocks commenced early in the morning; the first was followed at an interval of an hour by a second, and then by others at shorter intervals and with increasing violence, until at one o'clock P.M. a very severe shock was felt all through the southwest part of the island. From this time until the 10th of April the earth was in an almost constant tremor. In the district of Kona as many as fifty or sixty distinct shocks were counted in one day; in Ka'ū over three hundred in the same time. . . . It is said that during the early part of April two thousand distinct shocks occurred in Ka'ū, or an average of one hundred and forty or more each day. . . .

Every stone wall, almost every house, in Ka'ū was overturned, and the whole was done in an instant. A gentleman riding found his horse lying flat under him before he could think of the cause, and persons were thrown to the ground in an equally unexpected manner. (Brigham 1909, 101)

The violent quake evidently triggered slippage along many other faults, allowing the southern flanks of Mauna Loa and Kīlauea to slump seaward. Much of Kīlauea's southern coast sank into the sea during the earthquake, causing a tsunami that destroyed coastal villages and drowned forty-six people between Ka Lae and Cape Kumukahi. Aftershocks continued for more than eight months.

South Point (Ka Lae)

Six miles west of Wai'ōhinu, Hawai'i 11 intersects South Point Road, which leads 9 miles to a dead end at South Point (Ka Lae), the southernmost tip of land in the United States. The yellow soils around this windy headland are the weathered ash deposits of many eruptions from Kīlauea, 40 miles away. They derive their color from a stain of iron oxide.

Look inland from Ka Lae to see the broad shield of Mauna Loa. The remarkably flat skyline is not the summit, but a prominent shoulder on the slope of the volcano at about 12,000 feet. Geologist Robin Holcomb believes

Weathered basalt from a Mauna Loa eruption about 0.5 miles northeast of South Point. Like most beaches along this shore, the basalt contains abundant olivine. Some of the beaches along this stretch of coastline are green with olivine.

101

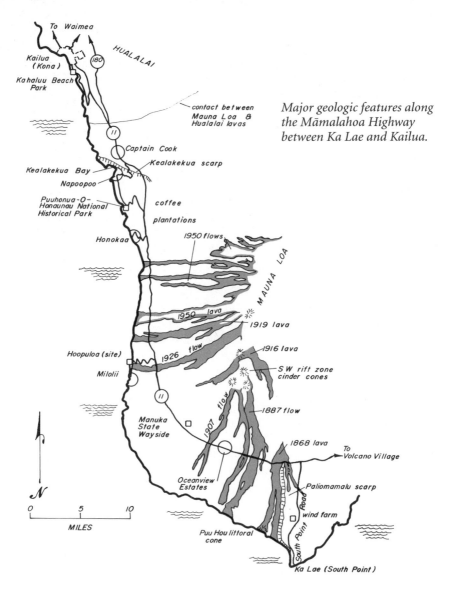

To Waimea

HUALALAI

Kailua (Kona)

Kahaluu Beach Park

180

11

contact between Mauna Loa & Hualalai lavas

Major geologic features along the Māmalahoa Highway between Ka Lae and Kailua.

Captain Cook

Kealakekua scarp

Kealakekua Bay

Napoopoo

Puuhonua-O-Honaunau National Historical Park

coffee

plantations

Honokaa

1950 flows

MAUNA LOA

1950 lava

1919 lava

1916 lava

Hoopuloa (site)

1926 flow

SW rift zone cinder cones

Miloli'i

11

1907 flow

1887 flow

Manuka State Wayside

1868 lava

To Volcano Village

N

Oceanview Estates

Paliomamalu scarp

0   5   10

MILES

wind farm

South Point Road

Puu Hou littoral cone

Ka Lae (South Point)

this shoulder could have formed when a former caldera of Mauna Loa, much larger than Mokuʻāweoweo caldera, was filled to overflowing with lava. The young flows then built the modern summit, leaving a break in the slope that marks the edge of the old caldera. Mokuʻāweoweo formed later, around 600 years ago.

The high sea cliff that continues inland west of Ka Lae is Pali o Kūlani, the slump scarp where the southwest rift zone dropped. It continues north as Paliomāmalu. The scarp appeared when a large submarine slide broke across the submerged slope of Mauna Loa and spread into the Hawaiian Deep 40 miles to the south. Most of the rocks exposed in the face of Pali o Kūlani are thin flows of ʻaʻā basalt, the dominant lava on the steep, lower flanks of Mauna Loa.

The core of an 'a'ā flow lies exposed along Hawai'i 11 in the southwest rift zone of Mauna Loa. The vertical columns formed when the shrinking lava crystallized.

In 1868, a huge 'a'ā flow erupted from the base of the scarp. You can see it entering the sea to the west. A large littoral cone, Pu'u Hou, marks the point where one tongue of lava poured into the ocean. Wave erosion has cut the cone in half.

Southwest Rift Zone of Mauna Loa

The core of an 'a'ā flow that crops out along Hawai'i 11 west of the intersection with South Point Road displays a beautiful row of basalt columns. Farther west, the highway crosses flows that erupted from the southwest rift zone in 1868, 1887, and 1907.

Watch for a scenic overlook about 5 miles west of the South Point intersection. It provides a spectacular view of the huge slide scarp and of Ka Lae. Upslope, you can see numerous cones and flows in the southwest rift zone.

The overlook is on lava that erupted in 1907. A nearby roadcut exposes the massive flow core, with oxidized top and bottom breccias and a rich abundance of green olivine crystals. About a third of the subdivision at Oceanview, west of the overlook, is laid out on this flow and on an earlier one that erupted in 1887.

West of Oceanview, Hawai'i 11 enters the district of South Kona and begins winding north along the steep western flank of Mauna Loa, crossing several recent lava flows.

The highway reaches the 1926 flow about 6 miles north of Manukā State Wayside, just north of the macadamia nut orchards. This lava reached the shore below and destroyed the village of Ho'ōpūloa.

Miloliʻi and the 1950 Eruption

A side road winds 5 miles down to the coast at the site of Hoʻōpūloa and the village of Miloliʻi.

The residents of Miloliʻi depend heavily on fishing. Next to the boat ramp in the center of the village is Miloliʻi Beach Park. A small seawall separates a picnic area and playground from a rocky shore, with shallow pools well suited to swimmers and snorkelers.

About a mile south of Miloliʻi, a walk along a rough road and trail takes you to Honomalino Bay, and to one of the Big Island's largest and prettiest beaches, which is privately owned. The sand is gray because it is a mixture of bits of glassy black lava, crystals of green olivine, and pulverized coral.

A few miles past the Miloliʻi turnoff, the highway crosses the 1919 ʻAlikā lava flow. Then, in the 6 miles or so south of Hoʻokena, it crosses three tongues of ʻaʻā lava that erupted in 1950. The long, steep slope below the highway is Pali Kaholo, the top of a large slump in the lower western flank of Mauna Loa.

The spectacular and threatening 1950 eruption began when a fissure about 15 miles long opened from Mokuʻāweoweo down the southwest rift zone toward the sea. About 580,000,000 cubic yards of lava erupted in a few days, an amount comparable to several years' production in the recent eruptions at Kīlauea. A dazzling curtain of fiery lava rose in places to heights of a thousand feet. Most of this lava spilled down the steep western flank of the volcano into the nearby South Kona district, which has many plantations and villages. Flows moved down the steep slope as fast as 6 miles an hour, even where dense forest stood in the way. In the dark hours of the first night, lava cut across Māmalahoa Highway, buried a community from which many people barely escaped, and poured into the sea.

Not far past the 1950 flows, the highway passes a side road leading to Hoʻokena. The lane winds 2.3 miles from Hawaiʻi 11 to the coast. Near the end of the road, on rocky Kauhakō Bay, is a small gray beach of mixed lava and coral sand, complete with a coconut grove for shade. The water here deepens quickly offshore; for most of the year, it is calm. Colorful fish attract snorkelers. This coastline is too young and sinking too fast to have developed a wide reef, but corals have made a patchy veneer across the bottom.

Puʻuhonua o Hōnaunau and Kealakekua Area

Farther north, a drive of 3.5 miles downslope from Hawaiʻi 11, on Hawaiʻi 160, takes you to the shore, and to Puʻuhonua o Hōnaunau National Historical Park. The living reef along much of this shore makes Hōnaunau Bay a popular snorkeling destination.

Roadcuts beside the scenic overlook above the coast near the park expose thin lobes of pāhoehoe lava. The huge cliff to the north, dotted with homes, is Pali Kapu o Keōua, a fault scarp that was the focus of a strong earthquake in 1951. The scarp probably originated about 100,000 to 105,000 years ago, at the head of the enormous 'Alikā landslide. The landslide may have generated the giant tsunami that scrubbed the soil off the island of Kaho'olawe to a height of 800 feet and washed blocks of coral even higher up the side of the island of Lāna'i. Kaho'olawe is 80 miles away; Lānai is farther.

Most of the coastal flatland between the scenic overlook and Pali Kapu o Keōua consists of pāhoehoe lava. The pāhoehoe caps a delta of lava and debris built up across the top of the 'Alikā Slide.

A visitor trail in Pu'uhonua o Hōnaunau loops through the compound of an ancient religious site built on a pāhoehoe flow that makes a sort of delta. Tsunami and storm waves moved the huge blocks of lava scattered across the flow surface near the shore, a few inside the compound. Some desperate day another tsunami will bring more to join them.

North along the shore from Pu'uhonua o Hōnauanan, you reach the village of Nāpo'opo'o. From the waterfront there, you can look across Kealakekua Bay to the imposing face of Pali Kapu o Keōua. Captain Cook visited this bay in 1778 and later died here because of a fight with a party of Hawaiians. A monument to him stands on the opposite shore.

A small eruption at a depth of 150 to 200 feet on the floor of Kealakekua Bay lasted about ten days in 1877. Its source was one of the many fissures that radiate across Mauna Loa's northwestern flank. Landsliding may have opened fractures in the volcano, providing easy passage for the rising magma.

Following Hawai'i 11 north from the Kealakekua Bay area, the road divides at Honalo, near the contact between Mauna Loa and Hualālai. Hawai'i 11 descends 1,300 feet to Kailua. The large mountain inland from Kailua is Hualālai.

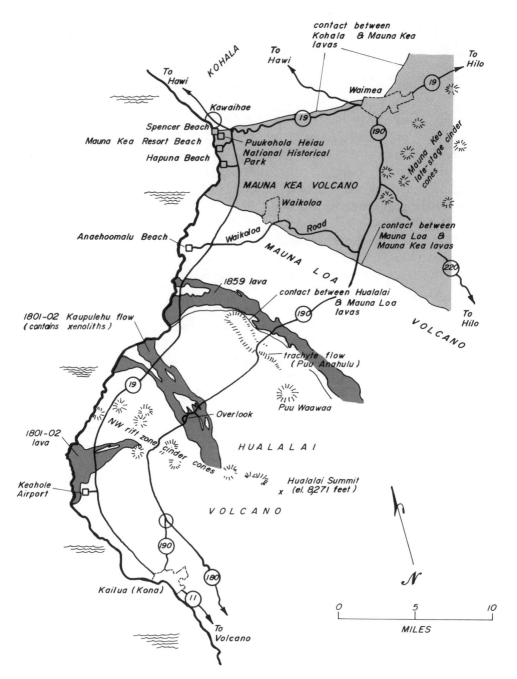

Major geologic features along Hawai'i 19 and 190 between Kailua and Waimea.

Hala (pandanus) growing on pāhoehoe lava. The roots of this common coastal plant effectively break down lava.

Hawai'i 19, Queen Ka'ahumanu Highway
Kailua–Waimea
40 miles

Queen Ka'ahumanu Highway follows the coastal route between Kailua and Waimea. It crosses young Hualālai and Mauna Loa lava flows, passes some of the state's finest beach parks, and ends at the foot of Kohala, oldest of the island's five roadside volcanoes.

Two of the Big Island's most popular swimming areas, Kahalu'u and White Sands beach parks, lie along Ali'i Drive, a few miles south of Kailua-Kona center. Kahalu'u Beach is a short, wide stretch of white calcareous sand with a fringe of coconut palms. A sheltering reef continues inshore almost to the beach. The water is just deep and calm enough for excellent swimming and snorkeling. White Sands, sometimes called Magic Sands, is more exposed to the open ocean than Kahalu'u Beach. A small sandbar

Reef development around the Big Island is sparse, but inshore waters are rich in marine life. —National Park Service photo

shelters the inshore water, but high seas sometimes sweep across it, causing dangerous rip currents.

North of Keāhole Airport, Hawaiʻi 19 crosses the coastal portion of Hualālai's northwest rift zone. The lava north of the airport is from the flow of 1800 and 1801, the latest eruption. The flow added new land to about 15 miles of coastline, burying many fishing villages and filling a bay in the process.

Reverend William Ellis compiled an account of the eruption from interviews with witnesses:

> Stone walls, trees, and houses, all gave way [before the flow], even large masses of hard, ancient lava, when surrounded by the fiery stream, soon split into small fragments, and, falling into the burning mass, appeared to melt again, as borne by it down the mountain's side. Offerings were presented [by the natives], and many hogs thrown alive into the stream to appease the anger of the gods, by whom they supposed it was directed, and to stay its devastating course. All seemed unavailing, until one day the king Kamehameha, went attended by a large retinue of chiefs and priests, and, as the most valuable offering he could make, cut off part of his own hair, which was always considered sacred, and threw it into the torrent. A day or two after, the lava ceased to flow. The gods, it was thought, were satisfied. (Brigham 1909, 14)

Like most young lava flows on Hualālai, this lava is alkalic basalt that erupted during the late stage of volcanic activity. Many cinder cones upslope mark the rift zone to the summit of Hualālai. The housing developments on steep slopes close to this active rift zone are situated in considerable volcanic danger.

The place where Mauna Loa and Hualālai lava flows meet is about 10 miles north of Keāhole Airport, just north of a prominent bend in the highway. Hawai'i 19 crosses a young pāhoehoe flow that erupted from the upper flank of Mauna Loa in 1859. The lava traveled 25 miles in three hundred days.

Look inland from this stretch of road to see the high Humu'ula Saddle between Mauna Loa and Mauna Kea. The prominent cinder cone green with plants on the flank of Hualālai is Pu'u Wa'awa'a. Directly ahead rises Kohala, the northernmost portion of the Big Island.

'Anaeho'omalu Beach

Hawai'i 19 meets Waikoloa Road about 25 miles north of Kailua. Turn west toward the shore to reach 'Anaeho'omalu Bay and a beautiful, palm-shaded beach. The white sand comes almost entirely from the reefs. The beach slope offshore is sandy and gradual, unlike most beaches on the Big Island.

The rough 'a'ā lava on the south side of the bay, which is full of caves and inlets, erupted from Mauna Loa. On a clear day, you can look inland to see the summits of Hualālai, Mauna Loa, Mauna Kea, and Kohala.

Three to four miles north of Waikoloa Road, the highway passes onto hummocky and sparsely vegetated Mauna Kea lava flows. The hummocky terrain is typical of land underlain by hawaiite, a common type of alkalic lava that typically erupts during late-stage volcanic activity.

Hāpuna and Spencer Beaches

Hawai'i 19 meets the road to Hāpuna about 30 miles north of Kailua. Like 'Anaeho'omalu, Hāpuna Beach is made of tan calcareous sand eroded from offshore reefs. The beach has a gradual submarine slope, making it ideal for swimming and wading on calm days. On clear days, you can see the crest of Haleakalā on Maui, 10,000 feet high and 50 miles to the northwest, across 'Alenuihāhā Channel.

A rare variety of basaltic lava called ankaramite makes up the ledges at the north end of the beach. This flow erupted from Mauna Kea during late-stage activity. Look for blocky crystals of black pyroxene and glassy crystals of yellowish green olivine.

North of Hāpuna, Hawai'i 19 turns east and climbs inland toward Waimea. A half mile drive west on Hawai'i 270 will take you to Samuel M. Spencer Beach Park and Pu'ukoholā Heiau National Historical Site.

The beach at Spencer Park is a wide reach of gently sloping calcareous sand, much like the beaches at Hāpuna and 'Anaeho'omalu. Harbor works and an offshore reef provide better shelter than at other North Kona beaches. The swimming and snorkeling here are excellent.

109

Puʻukoholā Heiau, the remains of a large sacrificial temple to the war god Kūkāʻilimoku, overlooks Spencer Park. Kamehameha was a patron of this heiau throughout his conquests.

Hawaiʻi 19 climbs to Waimea south of the contact point between lava flows from Kohala and Mauna Kea. Most days, Waikoloa Stream, south of the highway, is full with runoff from the rainy Kohala summit area, a stark contrast to the surrounding semiarid landscape.

Hawaiʻi 190, Māmalahoa Highway
Kailua–Waimea
50 miles

Hawaiʻi 190 rounds the northwest rift zone of Hualālai about 9 miles north of Kailua, just north of the Kalaoa subdivision. A scenic lookout west of the road stands on a lava flow that erupted in 1800 and 1801. This flow is noteworthy for the remarkable abundance and variety of xenoliths, including fragments of dunite, gabbro, and peridotite. Look for the angular chunks of dark to light green rock embedded in the dark gray lava.

About 8 miles north of the scenic overlook, Hawaiʻi 190 makes its closest approach to Puʻu Waʻawaʻa, the large trachyte cinder and pumice cone about 2 miles south of the road. It erupted about 100,000 years ago, producing a

Puʻu Waʻawaʻa. The ridge slope at the left is part of a thick trachyte flow that erupted from this vent.

set of thick trachyte flows. Unburied portions of these flows form the low ridge around which the highway curves.

Hawai'i 190 leaves Hualālai about 1.5 miles north of the trachyte flows and passes onto Mauna Loa. The youngest of the Mauna Loa flows is the pāhoehoe lava of 1859, which continues to the coast below.

About 5 miles farther north, the highway crosses older and more eroded flows that erupted from Mauna Kea. Watch for excellent views of Maui, in the distance to the northwest.

Many cinder cones that erupted during the late stage of volcanic activity in the western rift zone of Mauna Kea lie upslope where the highway approaches the saddle between Mauna Kea and Kohala. Look north to see the town of Waimea, and Kohala's eroded volcanic shield. Like Mauna Kea, it is crowned with cinder cones that grew during late-stage activity.

Hawai'i 250
Waimea–Hāwī
20 miles

North of Waimea, Hawai'i 250 climbs high up the flank of Kohala, skirting the volcano's summit area. For most of the route, it follows the northwest rift zone, passing cinder cones from late-stage volcanism.

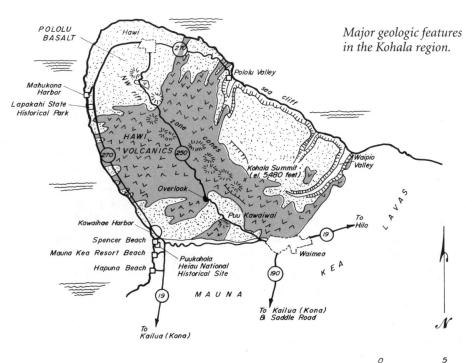

Major geologic features in the Kohala region.

A quarry exposes the core of Puʻu Kawaiwai.

About 2.5 miles north of Waimea, the road passes above Puʻu Kawaiwai, a small cinder cone with a quarry in its side. It consists of the common alkalic lava hawaiite.

Look in the quarry wall nearest the road to see the profile of a crater that was buried as the cinder cone grew. Spatter and cinders filled the crater when the eruption shifted to three other vents farther down the slope. Several lava bombs are embedded in the quarry walls and scattered across the surrounding ground. Puʻu Kawaiwai derives its red color from an iron oxide stain that formed as hot gases, mainly steam, oxidized loose cinder inside the cone.

A Dome of Rare Rock

A scenic overlook about 2 miles north of Puʻu Kawaiwai is almost on the axis of Kohala's northwest rift zone. The view south reveals Hualālai and the north Kona coast. The huge bulk of Mauna Loa rises above Hualālai to the east. You can see the full lengths of the flow of 1859 from Mauna Loa, and the flow of 1800 and 1801 from Hualālai, both of which reach the coast.

A dome of an extremely unusual type of alkalic lava called benmoreite, and some hardened rock rubble, are exposed in a roadcut across the street from the overlook. Benmoreite is so rare that most geologists have never heard of it. Its most distinctive feature is large crystals of black amphibole. You rarely see amphibole in Hawaiʻi because few Hawaiian magmas contain the ingredients for making it. You can distinguish amphibole crystals from pyroxene because they tend to be long instead of blocky and have glossy instead of dull surfaces. The rock also contains large crystals of pale plagioclase, as well as scattered small crystals of olivine weathered to rusty specks.

A bit less than a mile up the slope is Puʻu Loa, one of the youngest cones on Kohala. It erupted 120,000 years ago. The lava flows from it surround the benmoreite and also make up the terrain around Puʻu Kawaiwai. The highway passes other cinder cones as it follows the axis of the northwest rift zone down the slope to Hāwī.

Hawai'i 270
Pololū–Kawaihae
30 miles

Hawai'i 250 ends in Hāwī at the Hawai'i 270 intersection. Hawai'i 270 winds east to an overlook at the rim of Pololū Valley, where you can see a spectacular panorama of the huge sea cliff on the north shore of Kohala.

The big cliff along Kohala's northern coast is the headwall of a slide that carried debris northeast across the ocean floor as far as the Hawaiian Deep, 75 miles away. The slide must have had tremendous momentum to move so far on such a gentle slope; it probably caused an enormous tsunami. Sea stacks show that waves have eroded the cliff face at least a short distance inland. Waterfalls abound.

The trail from the parking area leads about half a mile to the bottom of Pololū Valley. The bedrock valley walls extend downward in a much deeper canyon that was flooded as the island slowly sank and when sea level rapidly rose at the end of the most recent ice age. Streams poured sediment into the submerged part of the canyon, filling it and creating the flat valley floor that you see today. These sedimentary deposits consist mainly of muds, sands, and gravels washed down from weathered lava flows during periods of heavy rain. Peat beds and small coal seams may also lie beneath the valley floor.

The devastating tsunami of April 1946 struck Pololū Valley especially hard. First, the water withdrew from the shore, exposing broad bedrock flats on the wave-cut bench. Then it surged back, rising to 55 feet as it swept across the beach at the mouth of the valley.

At the head of Pololū Valley, about 4 miles inland, lava flows that erupted during the late stage of volcanic activity (about 140,000 years ago) poured 2,500 feet down steep cliffs, accumulating at the bottom to form a nearly level valley floor. Elsewhere, the valley walls are mainly shield basalt flows from the older Pololū formation.

West of Hāwī, Hawai'i 270 follows the coast of Kohala toward Kawaihae. The change in vegetation between the dry western slope and wet northern slope is striking. So is the contrast between the deeply eroded and collapsed northern flank of Kohala and the barely eroded western slope.

One reason the western side of the volcano is more erosionally stable is because a large undersea shield, Māhukona, lies 30 miles off this coast, providing flank support. Māhukona once rose 700 to 800 feet above sea level, before isostatic sinking dropped its summit beneath the waves.

Māhukona and Lapakahi Parks

About 6.5 miles west of Hāwī, Hawai'i 270 passes the entrance to Māhukona Beach Park. This is hardly a beach at all, but an abandoned port for the Kohala Sugar Company. World War II and the 1946 tsunami put the port out of business. The harbor is now teeming with colorful marine life, much to the delight of snorkelers and divers. Nearby, Lapakahi State Park, inhabited by Hawaiians until the mid-nineteenth century, is now an archaeo- logical center with many partially restored sites.

Koaiʻe Cove has many sea caves and an abundance of marine life. Most of the caves are lava tubes.

A long roadcut between 0.7 and 1.5 miles south of Lapakahi State His- torical Park exposes a basalt flow somewhat enriched in sodium, but with fewer dark minerals than alkalic basalt. Technically, it is mugearite.

The Hāmākua Coast, Hawaiʻi 19
Waimea – Hilo
45 miles

The saddle pastureland between Mauna Kea and Kohala is one of the prettiest places in the Hawaiian Islands. Remarkable changes in plant cover accompany the change from dry country on the west side to wet on the east side. The forested east rift zone of Kohala lies along the horizon to the north, and Mauna Kea dominates the horizon to the south. The dozens of

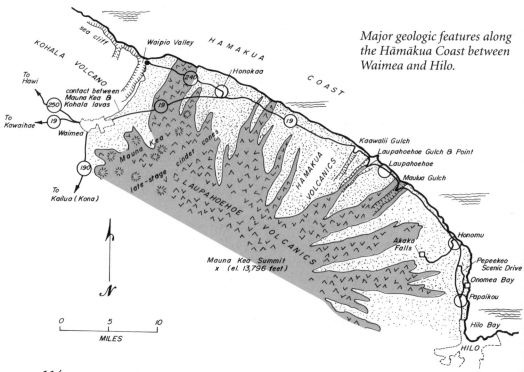

Major geologic features along the Hāmākua Coast between Waimea and Hilo.

Laupāhoehoe alkalic cinder cones scattered across Mauna Kea erupted over the past 65,000 years. The smooth, pale gray surface around the summit of Mauna Kea is glacial deposits. On clear days, you can see the summit astronomical observatories, 15 to 20 miles away.

Waipiʻo Valley

Hawaiʻi 240 heads east from Hawaiʻi 19 at Honokaʻa, following the coast for about 8.5 miles to Waipiʻo Valley Overlook. Like Pololū Valley near Hāwī, Waipiʻo Valley was once a much deeper canyon. It too is gradually sinking, and Waipiʻo Stream has covered the valley with sediment, creating a broad valley floor. Early Hawaiians heavily cultivated this verdant floodplain, which was an important population center in old Hawaiʻi.

Rocks exposed in the walls of Waipiʻo Valley are slightly alkalic lavas in the upper part of the Pololū formation. They erupted during Kohala's final activity. Many lava flows are chemically typical of shield basalts but contain spectacular crystals unlike any that formed during the main period of activity.

A walk along the road to the floor of Waipiʻo Valley reveals some interesting rocks. Several of the lava flows exposed in the roadcuts consist of large crystals of white plagioclase, some as much as an inch long. They are weathered, soft, and crumbly. Such a concentration of large crystals can accumulate as they float near the top of a magma chamber that is stagnant for a long time. Eruptions are too frequent during the main phase of shield growth for that to take place. So it seems likely that the plagioclase concentrated while eruptive activity was waning, toward the end of the shield-building stage.

After you reach the valley floor, look in the nearby stream bed or on the beach for cobbles of lava that contain many large crystals of black pyroxene and green olivine.

Lālākea Stream enters Waipiʻo Valley across two narrow waterfalls, each about 300 feet high. You can see these beautiful falls by looking straight ahead from the point where the road reaches the valley floor. A lava flow from Mauna Kea, part of the Hāmākua formation, poured down the slope east of the falls, following a stream that has since carved a new channel across the lava flow. This is one of the few flows that poured into Waipiʻo Valley. The others are lobes of lava from the Hāwī formation that erupted from the summit of Kohala.

The mouth of Waipiʻo Valley acts as a funnel for tsunami, raising the giant waves to towering heights. The 1946 tsunami crested at 40 feet in Waipiʻo Valley and swept inland for half a mile.

Hāmākua Flank of Mauna Kea

Between Honokaʻa and Hilo, Hawaiʻi 19 crosses the slopes of Mauna Kea's extinct shield. The rocks are mainly basalt flows in the Hāmākua formation. Deposits of volcanic ash in the Laupāhoehoe formation erupted from vents

115

near the summit of Mauna Kea. They are 15 feet thick on much of this older landscape, and weather into soils that nourish large fields of sugarcane. Geologically young gulches, some quite large, cut through the shield flank.

Not quite 12 miles east of Honoka'a, Hawai'i 19 curves down through Kawāili Gulch. The large roadcut at the northern rim exposes a flow of Laupāhoehoe alkalic basalt. Across the middle slopes of the gulch, the road slices across unusual alkalic basalts of the Hāmākua formation, erupted during the sluggish late stages of shield volcanism. Some of these flows are ankaramites rich in large crystals of black pyroxene and green olivine. The lower slopes are more ordinary flows of olivine tholeiitic basalt that erupted when Mauna Kea's shield was growing rapidly.

Watch for Laupāhoehoe Overlook on the seaward side of the highway south of Kawāili Stream and north of Maulua Gulch. It provides a view of Laupāhoehoe Peninsula, a late flow that erupted from a vent well upslope, and poured down Laupāhoehoe Gulch. Below the overlook, a steep road descends the former sea cliff to the peninsula. From the flats below, you can look through Laupāhoehoe Gulch all the way to the summit of Mauna Kea.

The 1946 tsunami was 30 feet high at Laupāhoehoe. It swept across the peninsula, destroyed a schoolhouse, and drowned twenty-four people, most of them children. The water offshore is almost always far too rough for swimming.

A few miles south of Laupāhoehoe Gulch, Hawai'i 19 enters another large gulch that Maulua Stream carved to a depth of 900 feet. The oldest exposed rocks on Mauna Kea are on the floor of this gulch. They erupted 150,000 years ago, when the volcano was rapidly building its shield. They are part of to the Hāmākua formation.

A short distance south of Hakalau, Hawai'i 11 passes Kolekole Beach Park. A lovely park, it is at the bottom of the deep gulch. A rocky black sand beach is at the mouth of a wide stream, and a big estuarine pool and small tributary waterfall lie behind the beach bar.

Most of the black sand beaches on the Big Island were formed when molten lava shattered into fine particles of glass as it entered the ocean. The supply of glassy black sand is severely limited. The black sand at Kolekole Beach comes from slow stream erosion of basalt lava flows farther up the valley and from waves eroding rocks along the shore. These processes provide a continuing and reliable supply of black sand that will maintain Kolekole Beach indefinitely. The grains of black sand are nicely rounded and dull, not glassy.

The 1946 tsunami swept into Kolekole Gulch with a crest 40 feet high. The huge waves undermined the bridge over the park, which was part of the Hāmākua Coast Railway.

'Akaka Falls

At Honomū, four miles of side road wind up the slope, crossing cultivated slopes of red laterite weathered on Laupāhoehoe ash. The road ends at

'Akaka Falls State Park, on the stream gulch that flows to Kolekole Beach. 'Akaka Falls is 440 feet high. Kahūnā Falls, downstream, is 400 feet high. Both waterfalls tumble across resistant ledges of Hāmākua lava at the head of Kolekole Gulch. You can see scars from fresh avalanches on the oversteepened cliff faces above the plunge pools. Floods, which typically follow winter storms, wash away the avalanche debris.

A network of little waterfalls resembles delicate shreds of lace on the cliffs. Water soaks into the porous lava flows high on the slopes, percolates through them, and issues from the cliffs in pretty little springs and seeps.

Hawai'i 200 and 220, The Saddle Road
Hilo–Waimea
65 miles

The narrow, winding Saddle Road crosses the mainly unpopulated central part of the Big Island. It passes through Humu'ula Saddle, the gap separating Mauna Kea and Mauna Loa, at 6,500 feet.

The west end of Hawai'i 220 begins at the Hilo waterfront as Waiānuenue Avenue. It passes Rainbow Falls, Boiling Pots, and Kaūmana Cave as it climbs

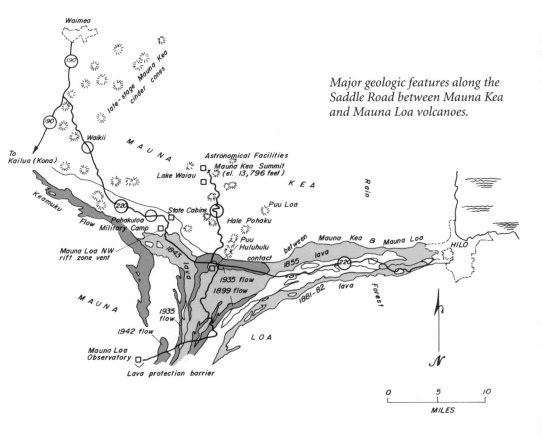

Major geologic features along the Saddle Road between Mauna Kea and Mauna Loa volcanoes.

through lush 'ōhi'a forest. The heavy annual rainfall and warm temperatures here favor some of the most rapid vegetation growth in Hawai'i. Much of this greenery thrives on lava flows that erupted from Mauna Loa in historic time.

For several miles past Kaūmana, the road gradually ascends a flow from 1881 and 1882. About 10 miles west of Hilo, the road passes onto older lava that erupted in 1855 and 1856. The obvious change in plant cover announces that rather small age difference.

Over the next 10 miles, on the way to the saddle, Hawai'i 220 crosses more lava that erupted in 1855 and 1856. The plant cover thins as the road climbs into cooler elevations, and it becomes apparent that the flow is indeed young. In some places, the flow encloses patches of older land, called kīpuka. These are easy to spot because they contain tall stands of old trees.

Watch for large blocks strewn amid the rubbly 'a'ā of the 1855–56 flow. Most are fragments of channel levees that formed somewhere up the slope, tore free, and rode the flow down the slope. Some were coated with fresh lava as they drifted along, becoming giant lava balls.

Bombs on the Flow

About 23 miles west of Hilo, the Saddle Road leaves the 1855–56 flow and crosses a narrow belt of woodland growing on lava about 1,500 years old, then passes onto an 'a'ā flow that erupted in 1935 and 1936. This flow was the target of the first attempt ever made to control molten lava by aerial bombardment. Thomas A. Jaggar, director of the Hawai'i Volcanoes Observatory, persuaded the U.S. Army Air Corps to drop 500-pound bombs at the mouth of a lava tube feeding the main flow channel at 8,800 feet. Geologist Harold Stearns flew with the bombing mission and later wrote about the event:

> [Dr. Jaggar] hoped [the bombing would] divert the lava stream to keep it on the upper slopes and stop its advance toward Hilo. I obtained a seat in one of the planes as an observer. Jaggar stood in the [Humu'ula] Saddle area below us with a transit to watch the operation. Most of the bombs fell too far from the channel to do any good, and a few didn't even explode, as I found out from a field inspection later. One bomb hit the tube and threw out clots of lava but did not divert the lava from its channel. The flow stopped two days later, and Jaggar believed that the bombing had stopped the flow. I am sure it was a coincidence, however, like the time, in 1881, when Princess Ruth Keelikolani tried to stop a lava flow from invading Hilo by throwing thirty red silk handkerchiefs and a bottle of brandy into the molten flow. Several old Hawaiians told me later that all who dropped bombs would die by fire because of arousing the wrath of Madame Pele, the goddess of volcanoes. Later, I learned that many did die by fire when their planes crashed. Maybe I did not bring down the wrath of Madame Pele because I was just an observer. (Stearns, 1983, 108)

Undiscouraged, the Army Air Corps again tried bombs in the spring of 1942, hoping to control a flow coming down Humu'ula Saddle from Mauna

The Saddle Road passes Puʻu Huluhulu at the summit of Humuʻula Saddle.

Loa. The idea was to blast a hole in the levee of the main flow channel, causing lava to spread out high on the slope instead of running down to Hilo. The bombers did blow a hole in the levee, but only a small amount of lava leaked out, and the eruption ended before another effort could be organized. That was the last time the military went to war with an active Hawaiian volcano.

Puʻu Huluhulu

The Saddle Road follows the flow of 1935 and 1936, which covers most of the plain in the Humuʻula Saddle. Watch closely to see it change from ʻaʻā to pāhoehoe as the road follows it upslope toward the source.

At the crest of Humuʻula Saddle, about 30 miles from Hilo, a small wooded cinder cone rises south of the highway, west of an intersection with a red road that leads toward Mauna Loa. This is Puʻu Huluhulu, one of Mauna Kea's many alkalic cinder cones, part of the Laupāhoehoe formation. Younger lava flows from Mauna Loa have surrounded the base of Puʻu Huluhulu, turning it into a kīpuka.

Look near the northwestern base of Puʻu Huluhulu for a stone wall that was partially buried in pāhoehoe lava from an eruption in 1935 and 1936. You can see the original height where it continues into a kīpuka of older, brownish pāhoehoe lava just west of the cinder cone. If you walk along the edge of the wall, you can see that it lies in a trough in the lava. It almost appears as if something shoved the wall down into the flow.

After the lava pooled in the saddle and a hard crust formed on the surface, molten lava pouring down the long slope of Mauna Loa continued to feed the flow beneath the crust, lifting and splitting the crust as though it were rising bread dough. You can judge the amount of the rise as you walk along

119

A stone wall, surrounded and almost covered by lava from a 1935–36 eruption.

the wall. In a few places the lava buried the wall, but in most places the flow was too thin and the crust was too stiff to shift across the top of the wall.

The quarry in the western flank of Puʻu Huluhulu exposes the interior structure of the cinder cone. You can see basalt dikes cutting through crudely layered or structureless black cinder. The cinder cone consists of late-stage alkalic debris erupted from Mauna Kea, but the dikes are tholeiite basalt from the surrounding Mauna Loa lavas.

How could flows from Mauna Loa lapping around the base of Puʻu Huluhulu inject dikes into the cinder cone? Evidently, the molten pāhoehoe lava was moving beneath a surface crust that confined it under considerable pressure for the long flow down the slope. The pressure drove sheets of molten lava into the loose cinder. The lava cooled to form the dikes.

The dirt road through the quarry continues to the summit of Puʻu Huluhulu, where you can see the partially buried stone wall extending for hundreds of feet. The many cinder cones scattered across the near flank of Mauna Kea erupted in the volcano's south rift zone during late-stage volcanic activity.

The group of cones a few miles north, near the radio tower, erupted between 40,000 and 20,000 years ago. The younger cinder cones that rise farther up the slope include Puʻu Kole, Puʻu Loa, and Puʻu Huikau. Ash and cinder erupting from these vents about 4,500 years ago fell on a wide area, covering a foot-deep layer of soil that appears in gully cuts across the highway.

The upper portion of the buried soil contains bits of charcoal from wood presumably burned during the eruption. Radiocarbon dates on them provide geologists with their best information about the ages of volcanic rocks that are less than about 40,000 years old.

Dates on lava flows and charcoal from the southern flank of Mauna Kea suggest that the frequency of Laupāhoehoe eruptions in the latest ice age were at a rate of at least six or seven per 10,000 years. About twice as many have taken place since that ice age ended, about 11,000 years ago. Most were in the south rift zone. The latest eruption of Mauna Kea, about 3,300 years ago, built a cluster of small cinder cones on the upper south flank at around the 11,000-foot level.

Look west of the summit of Mauna Kea to see several small stream canyons dissecting an older part of the slope. They terminate near the summit, where a featureless blanket of gray material carpets the highest part of the slope. This is debris from the glaciers that covered the top of Mauna Kea during ice ages.

Looking from Puʻu Huluhulu toward Mauna Loa, you can see spatter ramparts and small cones with steep sides, all in the northeast rift zone, east of the summit. Flows of diverse age stand out on the high, barren slopes of the mountain in a patchwork of color from light brown to dark gray: the younger the flow, the darker the color.

Mauna Loa's Far-Flung Vents

About 5 miles west of Puʻu Huluhulu, 35 miles west of Hilo, the Saddle Road crosses prehistoric Mauna Loa lava, then turns abruptly toward Mauna Kea, parallel to a spatter rampart. The agglutinated surface spatter has weathered to shades of yellow and brown, though roadcuts reveal the dark unweathered spatter beneath.

This relic of an eruption where lava fountained from a long fissure seems curiously out of place among the other vents along the highway, which are mainly cinder cones. The lava is typical shield basalt, not late-stage alkalic rock. It is only 3,000 years old, younger than the latest eruptions from Mauna Kea. The spatter rampart is aligned toward the summit of Mauna Loa, about 20 miles away. In fact, this is a distant vent of Mauna Loa. The magma must have risen through the older Mauna Kea flows beneath.

Look up the slope to see Pōhakuloa Gulch, a prominent canyon high on the flank of Mauna Kea. Near the top of the gulch a large moraine outlines the lower end of a tongue of ice that reached the gulch some 150,000 to 100,000 years ago.

Across Mauna Kea's High Western Flank

Just west of Pōhakuloa State Park and the adjoining military camps, the road passes a scattering of grassy Mauna Kea cinder cones. The much younger

121

A Mauna Loa spatter rampart at the foot of Mauna Kea, near Pōhakuloa State Park. The smooth, pale slopes near the summit of Mauna Kea are glacial deposits.

lava flows surrounding some of them erupted from Mauna Loa. The flow of dark lava from Mauna Loa closest to the highway is 300 years old.

From this stretch of highway, you can see Hualālai Volcano to the southwest. Like Mauna Kea, Hualālai wears a cap of cinder cones, which gives it a rough profile. More than 120 cones dot the volcano, most in the northwest and southeast rift zones. The southeast rift zone is on the horizon left of the summit.

Near milepost 42, the road crosses onto Mauna Kea lavas, rises to the divide of the western rift zone, and begins a long and scenic descent to Waimea. You can see fine panoramas of Hualālai and its satellitic trachyte cone, Puʻu Waʻawaʻa, the northwest Kona coast, Kohala Volcano, Waimea Saddle, the widespread cinder cone field on Mauna Kea's northwestern flank, and Haleakalā, 70 miles away on the island of Maui.

The road crosses a hummocky slope as it descends the mountainside. The gently rolling landscape developed as soft ash was erupted from Mauna Kea and draped across rough flows of alkalic basalt. The ash weathered into rich soils that sustain the grassland. Most of the area is owned by the Parker Ranch, one of the largest privately owned grazing properties in the United States.

Mauna Loa Observatory Road, West from Hilo
Saddle Road–Mauna Loa Observatory
17 miles

A red cinder road off the Saddle Road at the intersection near Puʻu Huluhulu winds 17.4 miles up the northern flank of Mauna Loa to the Weather Observatory at 11,000 feet. The road is paved, but much of it is narrow and steep. It provides good views of lava flows weathering in all climatic zones, from moist subtropical to alpine. It also provides a feel for the enormous scale of Mauna Loa. Few places are quite like the barren volcanic wilderness high on the flanks of this mountain.

Because few posted markers guide you along the Mauna Loa Road, set your odometer to zero at the Saddle Road intersection.

The Observatory Road passes from the flow of 1935 and 1936 onto prehistoric lavas 0.2 miles from the Saddle Road. Two miles farther along, it crosses an extremely rough lava channel with high levees. The huge blocks scattered across the 1935-36 flow farther downslope originated where levee walls collapsed into the lava stream.

At mile 5.4, the Observatory Road reaches the dark, sparsely vegetated flow from the 1899 eruption. It passes two kīpuka of prehistoric lava in this flow. At mile 7.1, the road rises off the 1899 lava onto a thicker flow that erupted in 1855 and 1856; the surface is transitional from pāhoehoe to ʻaʻā. At mile 9.7, the road passes back onto 1899 lava. In the next 1.5 miles, the road crosses several branches from different historic flows, ranging from 1843 to 1942. Many kīpuka of prehistoric lava are in these flows.

Despite the excellent exposure, it is difficult at this altitude to locate the contact points where one young flow meets another. The flows may be decades apart in age, but the plant cover is sparse and weathering slowly, making the flows virtually indistinguishable.

At mile 13.6, you can see a fissure and spatter rampart several hundred feet upslope. This fissure is one of many widely scattered vents arrayed radially from the summit across the northwest flank of Mauna Loa. Vent structures of the northeast rift zone rise on the horizon a few miles away.

Many small kīpuka of dramatic prehistoric entrail pāhoehoe appear between miles 15.7 and 17.0.

The panorama from the parking lot at the end of the public road is spectacular. To the north is Mauna Kea, with white astronomical domes perched on the summit. Kohala, and more distant Haleakalā on Maui, are in the northwest.

The Mauna Loa Observatory is a facility for studying the earth's atmosphere. Scientists there have measured the increasing carbon dioxide content of the atmosphere since 1957. Their data inspired much of the current concern about a potential global greenhouse effect. They have recently begun monitoring ozone loss.

The trail to Mauna Loa's summit begins at the parking lot where the public road ends. It leads 3 miles past the Observatory, straight uphill to the stark rim of Moku'āweoweo caldera, one of the most desolate landscapes you will ever see. It is worth walking a few hundred yards up the slope to see the diversion barriers that protect the observatory from lava flows. These are the first such structures in the United States.

Mauna Kea Summit Road
Saddle Road–Summit of Mauna Kea
15 miles

The Mauna Kea Summit Road winds very steeply past cinder cones and across late-stage flows of the south rift zone to 9,200 feet, where it levels off at Halepōhaku and the Ellison B. Onizuka Astronomical Complex. The drive is too much for some cars, and for some drivers. Zero your odometer at the intersection with the Saddle Road.

Two prominent cinder cones appear on the eastern skyline about 5.5 miles from the Saddle Road. Pu'u Kole, which erupted 4,500 years ago, appears to be the younger. In the foreground is an alkalic basalt flow that erupted about 5,300 years ago from the low mound about half a mile up the slope.

The road passes through the breached, horseshoe-shaped crater of Pu'u Kalepeamoa. The ridge west of the road is the eastern rim of the crater, where the trade winds piled cinder high to one side. The cinder contains many fragments of older rock, including black gabbro and green dunite.

The visitor center at Halepōhaku is on a smooth blanket of ash and cinder that looks perfectly fresh. Beyond Halepōhaku, the road climbs nearly 2,000 feet up a steep and cindery slope in a series of five switchbacks. At the top it skirts the eastern base of a prominent cinder cone, Pu'u Keonehehe'e. Look southeast for a deep pit crater, one of the few that still exist on Mauna Kea. The road then enters glaciated terrain.

The Glaciers of Mauna Kea

The glacial deposits on top of Mauna Kea consist of till, deposited directly from glacial ice, and outwash, deposited from glacial meltwater. "Drift" refers to till and outwash together.

Glacial outwash on Mauna Kea typically consists of light-gray silt, sand, and gravel in neat layers. Outwash deposits in deeper canyons high on the mountainside also contain thick beds of chaotically mixed conglomerate.

Geologists interpret these deposits as flood dumps, perhaps the result of eruptions beneath Mauna Kea's former ice cap. If so, then at least six such subglacial eruptions took place. You can see an example of the more ordinary kind of outwash on the floor of the small valley near Puʻu Keoneh6heʻe, at mile 10.5.

Mauna Kea till is pale gray. It contains angular fragments of all sizes, randomly mixed. The glacial till and outwash deposits on Mauna Kea record four episodes of glaciation. The deposits include the Pōhakuloa drift laid down 150,000 to 100,000 years ago, the Waihou drift accumulated between 100,000 and 55,000 years ago, and the Makanaka drift deposited from 55,000 to 20,000 years ago. Except perhaps for the Pōhakuloa drift, each of these episodes corresponds to the Wisconsinan ice age of North America, which lasted from about 125,000 until about 11,000 years ago.

Most of the till along the road is the Makanaka drift. Scattered pale gray till drapes the dark gray or brown slopes of cinder cones farther up the slope. Glaciers eroded some of these cones almost beyond recognition.

Watch at mile 11.6 for scattered piles of basalt rubble. Ancient Hawaiians left them as they quarried the glassy lava to secure raw material making adzes. Federal and state laws strictly protect such sites.

Puʻu Waiʻau is about a mile farther northwest. Steam and hot water percolating through the cone near the end of its eruption altered the cinder, creating the light-colored patches. The alteration products include clay, which makes the rock impermeable, thus increasing runoff from rain and melting snow. So the flanks of Puʻu Waiʻau have more gullies than the more permeable flanks of neighboring, unaltered cones. Northwest of Puʻu Waiʻau is Puʻu Poliʻahu, another noticeably altered cinder cone.

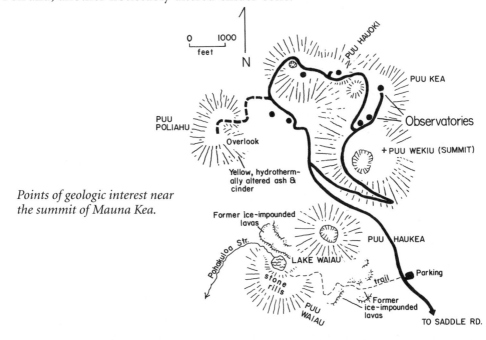

Points of geologic interest near the summit of Mauna Kea.

The Flow That Erupted under a Glacier

At mile 12.1, the summit cone of Mauna Kea, Puʻu Wēkiu, comes into view, with observatory domes perched near the top. For the next mile or so, the road follows a basalt flow that erupted from Puʻu Wēkiu. In many places, long tracks of parallel lines and mosaic zones of fracturing mark its surface. These suggest that the lava flowed beneath the ice, melting it and wedging its way along. The overlying ice created the peculiar fractures as it quenched the flow. Then it dragged the particles of grit across the lava, polishing its surface and etching long grooves and scratches.

Lake Waiʻau

A small parking area at mile 12.5 marks the beginning of a dirt path that leads half a mile west into the pass between two cinder cones, Puʻu Waiʻau to the south and Puʻu Hau Kea to the north. The path leads up the slope from the steep, lobate edge of a lava flow that erupted from Puʻu Hau Kea 40,000 years ago. The peculiar fracture patterns along the edge of the flow suggest that the cooling lava banked up against ice.

A few hundred yards beyond the pass lies Lake Waiʻau, at the bottom of Puʻu Waiʻau crater. It is one of the few natural bodies of fresh water in the Islands and, at 13,160 feet, is the highest lake in Hawaiʻi. The cinder piled high along the southern rim of the crater tells of strong winds from the north as the cinder cone grew. An extinct rock glacier occupies the crater just south of the lake. You can recognize this mixture of rock, once imbedded in ice and flowing toward the lake, as a pad of hummocky, light gray debris.

If the bottom of Lake Waiʻau were unaltered cinder, it would not hold water. But the floor of the crater contains beds of impermeable clay weathered from ash that erupted from Mauna Kea 3,300 years ago. Circulating hot water and steam may have helped by altering the cinder beneath the ash. Lake Waiʻau occasionally overflows through the notch in the western rim of the crater.

The embankment of rough lava along the north side of Lake Waiʻau is part of the flow that erupted from Puʻu Hau Kea. It poured across the low northern rim of the crater, then stopped. The cavernous voids, mosaic fractures, and lava pillows suggest that it stopped against ice. Look for many inclusions of dark and coarsely granular gabbro and green dunite, which consists mainly of olivine.

When light is low on the cindery slopes, you can also look for coarse fragments of cinder aligned in neat, evenly spaced rows a few inches apart, all stretching downslope. These cover many acres of the subarctic summit region of Mauna Kea. Just how they form is a mystery. Some geologists suggest that seasonal freezing and thawing of water in the shallow ground somehow sorts the cinder into neat rows.

126

Mauna Kea Summit

The road ends at the rim of Puʻu Wēkiu. Spindly bombs and blocks embedded in cinder beds are exposed in roadcuts along the way. The summit is on the high eastern rim of Puʻu Wēkiu crater, about 600 feet southeast of the parking area. A dozen astronomical observatories are a striking addition to summit scenery. Public tours of the observatories are offered occasionally. The night sky viewed from the summit of Mauna Kea is magnificent.

The view south from the parking area reveals the beveled summit of Mauna Loa, Mokuʻāweoweo caldera. To the southwest, you can see Hualālai, and to the north, Kohala. The misty blue profile of Haleakalā, on Maui, rises out of the ocean beyond. At sunset, the afternoon clouds that generally hide the Hāmākua and Hilo sides of Mauna Kea provide a striking base for an optical effect that looks like a huge, pyramidal shadow cast by Mauna Kea.

A volcanic bomb embedded in cinder near the summit of Mauna Kea.

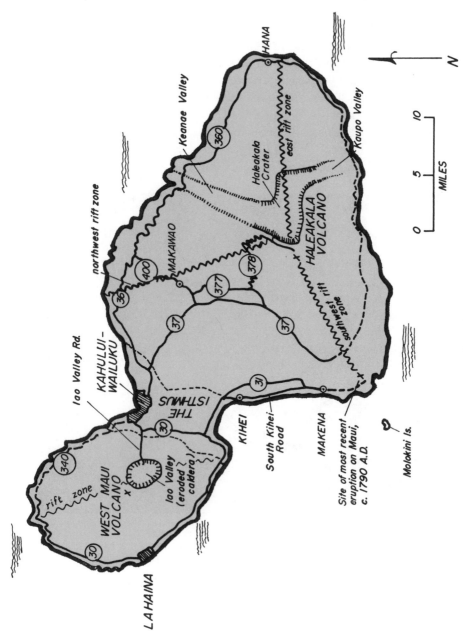

Major geologic features of Maui.

3
Maui, The Valley Isle

Maui is the second youngest of the main Hawaiian Islands. It consists of two large volcanoes, West Maui and Haleakalā, linked by the narrow Isthmus of Maui. West Maui seems to be extinct. Haleakalā erupted around 1790 and, although in its final decline, is almost certainly capable of erupting again. The landscapes of Maui range from subalpine volcanic terrain to lush tropical valleys, and include some of the finest beaches in Hawai'i.

During the latest ice age, when sea level was several hundred feet lower than it is now, Maui was part of a much larger island that geologists call Maui Nui. It was more than half the size of the Big Island. As ice melted at the end of the ice age, sea level rose, and the low areas of Maui Nui became the shallow straits that now separate the Maui group of islands: Maui, Moloka'i, Lāna'i, and Kaho'olawe.

West Maui Volcano

The older of Maui's two volcanoes, West Maui, rises 5,788 feet above sea level. It is 18 miles long and 15 miles wide. The oldest lava exposed above sea level on West Maui erupted nearly 2 million years ago, so the island must have been above sea level before then.

Thin flows of mainly pāhoehoe lava accumulated to build the young shield of West Maui. It was complete by 1.3 million years ago, with a caldera at the summit about 2 miles in diameter. The lavas erupted during this main stage of growth are the Wailuku basalts.

West Maui developed rift zones that trend north and south of the caldera. No third rift zone formed, perhaps because the older mass of neighboring East Moloka'i prevented movement in that direction.

In the usual way of Hawaiian volcanoes, the chemical composition of West Maui's lavas changed as volcanic activity declined. The frequent and generally mild eruptions of tholeiite basalt in the early stage of shield building gave way to more explosive eruptions of alkalic basalt and trachyte during late-stage volcanism. The new cinder cones and domes made the originally smooth profile of West Maui's shield look rough.

The late-stage rocks are the Honolua volcanic formation. They are not nearly as dark as the Wailuku basalts. If the weather is clear and the lighting

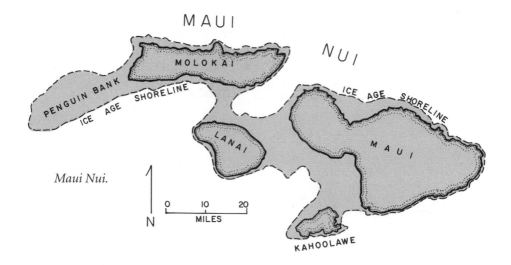

Maui Nui.

is right, you can easily distinguish the Wailuku and Honolua volcanic formations in the walls of the deep interior valleys on West Maui. The dark Wailuku basalts make up most of the lower slopes, and the light Honolua flows cover them, like pale frosting on a dark cake.

Surprisingly, West Maui's wet northeastern slopes are considerably less eroded than the dry southwestern slopes, even though rainfall and erosion rates are greater on the northeastern slopes. Because the younger Honolua volcanic rocks did not cover the dry southwestern slopes, they have been exposed to erosion far longer than the much younger terrain on the wet northeastern side.

The youngest Honolua lavas probably erupted about a half million years ago. Then rejuvenated volcanic activity built four cinder cones of highly alkalic basalt along the leeward shore, between Lahaina and the isthmus between the volcanoes. Two of those cones diverted streams, so erosion of the older volcanic rocks must have been well under way when they erupted. These latest and poorly dated volcanic rocks are the Lahaina volcanic formation. No one can tell whether those will be the last eruptions on West Maui. A volcano that last erupted rejuvenated-stage rocks a half million years ago may be truly dead.

Alluvial fans extending seaward from the mouths of valleys on West Maui created the narrow, gently sloping plain along the mountain's leeward coast. The growing fans rest on the floor of the shallow channel between Maui and Lānaʻi.

Haleakalā

Haleakalā means House of the Sun. In the tradition of early Hawaiʻi, the demigod Māui used a net to snare the sun as it crossed the sky above Haleakalā, slowing it down so that his mother, Hina, could have time to dry her kapa cloth.

Haleakalā, 33 miles long and more than 20 miles wide, first rose above sea level around 900,000 years ago. At 10,025 feet, the summit is high enough to catch some snow. Haleakalā is not as high as Mauna Kea and Mauna Loa, on the Big Island, but it is older than they are and has sunk much farther into the ocean floor. If you consider the entire volcano, the parts above and below sea level, Haleakalā is the largest volcano in the Hawaiian Islands.

In its prime, Haleakalā was a vast shield of olivine tholeiite basalt lava flows. Geologists call these early Haleakalā lavas the Honomanū basalts. They are analogous to the Wailuku basalts on West Maui, but younger. A succession of calderas probably formed at the summit of the growing shield, but younger rocks filled and buried them. About 700,000 years ago, shield growth slowed, and explosive eruptions began to produce more alkalic rocks. Haleakalā had entered the late stage of its development. Those eruptions, the Kula volcanic formation, continued until about 350,000 years ago.

By that time, the volcano had drifted north from the hot spot to approximately the present position of Mauna Kea. Haleakalā is now 140 miles northwest of the hot spot.

The rejuvenated stage of volcanism began on Haleakalā 100,000 years ago, and continues. Hundreds of cinder cones and flows of alkalic basalt with ʻaʻā surfaces have erupted during that time. Rejuvenated-stage volcanism has been much more intense than the comparable activity on West Maui. Rocks from the rejuvenated stage are the Hāna volcanic formation.

Haleakalā has active rift zones that trend southwest and east. The northwest rift zone is dead. The volcano slopes seaward about 5 to 10 degrees along that rift zone. The southern flank slopes seaward at an angle of more than 20 degrees, steeply enough to suggest that sliding may have contributed. But no evidence of a slide dump has been found on the adjacent ocean floor. In fact, Haleakalā must be tilting toward the Big Island because Hawaiʻi—a huge landmass—is sinking more rapidly than East Maui. That would also explain why ancient submerged reefs on the lower east rift zone of Haleakalā also tilt southward.

Lava pouring from the southwest and northwest rift zones of Haleakalā pooled against the flank of West Maui, building the isthmus that links the two volcanoes. Haleakalā then formed the broad, gently sloping lower flank along its western side.

The western flanks of Haleakalā are too young and too dry to have been much eroded. River valleys deeply dissect the wet eastern side, where thousands of cascades and waterfalls tumble down steep valley walls. Gently sloping remnants of the original flank of the shield separate eight huge valleys shaped like giant amphitheaters. Four of them meet at their heads, in an area that may have been the ancient caldera.

Ash, lava, and cinder erupted high in the Keʻanae and Kaupō valleys during rejuvenated volcanism, burying the low divide that separates the heads of these two drainages. The tops of the sheer cliffs around other parts of the valley heads stood above the new volcanic field, enclosing a vast, cindery, subalpine basin 7.5 miles long and 2.5 miles wide. At first glance this spectacular basin resembles a caldera, but it is primarily an erosional feature, though Keʻanae and Kaupō streams may have eroded it in the weak rock of some older caldera. Future eruptions may yet occur on the rugged floor of the basin.

In 1786, the French explorer La Pérouse mapped an embayment near the southern cape of the island, opposite Kahoʻolawe. In 1793, George Vancouver, sailing the same waters, found a peninsula of fresh lava entering the ocean where La Pérouse had mapped a bay. Accounts of a recent visit by Pele, the Hawaiian fire goddess, indicated that an eruption had recently taken place. This latest activity of Maui's giant House of the Sun will not be the last.

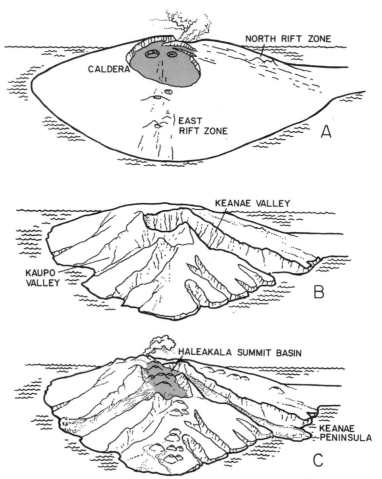

Haleakalā as it had declined from the maximum stage of shield growth 700,000 years ago to the present. The summit basin is not really a crater, even though it is in the approximate area of Haleakalā's possible ancient calderas.

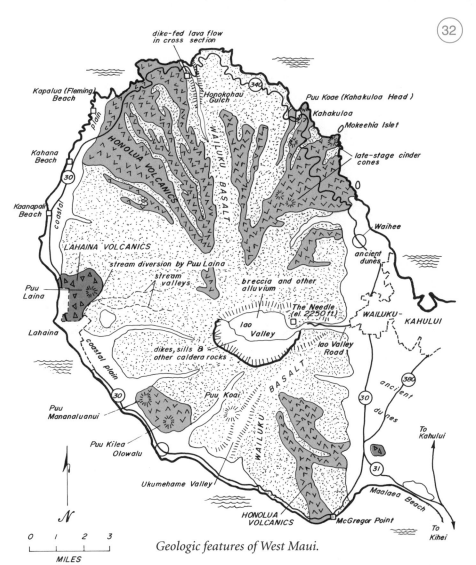

Geologic features of West Maui.

Hawai'i 32
Kahului–'Īao Valley State Park
3.5 miles

Ka'ahumanu Avenue passes through Kahului to Wailuku, where it intersects Hawai'i 30, the Honoapi'ilani Highway. Continue west to 'Īao Valley Road, Hawai'i 32 (320 on some maps), which heads from Wailuku into the spectacularly eroded caldera of West Maui. Zero your odometer at the intersection with Hawai'i 30.

A roadcut along 'Īao Valley Road 0.6 miles past the junction with Hawai'i 30 reveals a pebbly conglomerate, an alluvial fan deposit laid down in ancient floods. The nearby stream exposed the conglomerate as it eroded the valley.

'Īao Valley. Flank flows of Wailuku basalt make up the near ridge to the left. Breccias and caldera lava flows, greatly altered and full of dikes and sills, make up the farther valley walls.

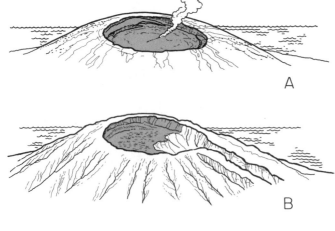

A

B

Erosion of West Maui's caldera produced 'Īao Valley. The Needle is one of many spectacular erosional features in the rock that encloses a natural amphitheater.

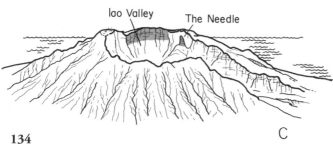

Iao Valley   The Needle

C

Thin flows of the Wailuku basalt that built most of the West Maui Volcano are exposed in the valley walls along the first 2 miles of 'Īao Valley Road. About 2.5 miles up the valley, you begin to see steeply dipping layers of weathered, rubbly rocks. Geologists interpret these as ancient landslide deposits at the foot of the steep wall that once faced into the caldera of West Maui. If you look upstream, you can see the cliffy rim of 'Īao Valley, which roughly corresponds to the old caldera.

Just 2.75 miles from the Hawai'i 30 junction, the road crosses Black Gorge, a deeply eroded tributary of the main stream that drains 'Īao Valley. Look upstream, low along the right side of the gorge, for the mass of rock that resembles a profile of President John F. Kennedy. It is a cylindrical boss of gabbro, a dark intrusive rock that has the chemical and mineral composition of basalt but in much larger crystals. The alkalic composition of this gabbro suggests that it dates from the period of Honolua volcanism. It may have crystallized beneath a pit crater similar to those in Kīlauea's east rift zone today.

'Īao Needle

The end of the road, almost 3 miles from Hawai'i 30, provides a view of 'Īao Needle. From this perspective, it looks like an isolated pinnacle. In fact, it is a knob standing on the crest of a winding ridge, the steep divide between two streams.

Rocks in 'Īao Needle are thin Wailuku basalt flows with many dikes cutting through them. Thicker lava flows, hardened rock rubble, and conglomerates that filled the caldera of West Maui appear in outcrops near the

'Īao Needle.

viewpoint and along ʻĪao Stream. They were deposited in the eastern part of the ancient caldera. The boundary lies somewhere between the visitor overlook and ʻĪao Needle.

In clear weather, you can trace the caldera boundary high along the valley wall southeast of the viewpoint. Look across the main valley opposite ʻĪao Needle. The old caldera wall appears as a steep boundary between thin basalt lava flows that dip to the northeast, to your left, and rocky debris that dips southwest, to your right. Sediments deposited in the caldera appear in the stream below the footbridge that leads to the shelter.

Rising magma drove hot water, steam, and other volcanic gases through the porous rubble and lava in the floor of the caldera. This caustic brew reacted with, and altered, the rocks. One of the reactions transformed the black mineral pyroxene into the green mineral chlorite, making the basalt green. Silica released in this reaction precipitated in fractures and vesicles as the minerals opal and chalcedony. Streams throughout ʻĪao Valley abrade and round fragments of the silica into smooth pebbles locally called moonstones. The altered rocks in the old caldera weather readily and erode easily, and that explains why ʻĪao Valley opens upstream into such a broad basin.

Hawaiʻi 30
Kahului–Lahaina
25 miles

Kahului is at the northern edge of the Isthmus of Maui. From this bustling little port, you can see towering Haleakalā, with its youthful shield still intact, and the green, deeply eroded flanks of West Maui, which are usually shrouded in clouds. Lava erupted from Haleakalā piled against the eastern flank of old West Maui Volcano to build most of the Isthmus of Maui. Alluvial fans and windblown dunes of calcareous sand, some as much as 200 feet deep, have since covered this narrow neck of land.

Broad expanses of offshore reef were exposed during the latest ice age, when sea level was several hundred feet lower than now. Strong winds blew calcite sand off the area around the exposed reefs and inland onto the isthmus. Today, plants stabilize most of the sand, but a few active dunes still shift about near the coast. Meanwhile, slightly acidic rainwater dissolves and reprecipitates the calcite, cementing the sand into solid limestone.

Between Wailuku and the intersection with Hawaiʻi 31, near milepost 5, Hawaiʻi 30 skirts the eastern base of West Maui, which is largely veneered with Honolua lavas. The small bumps on the flank of the volcano, on the horizon to the southwest, are eroded cinder cones, vents that erupted Honolua lavas along the south rift zone.

Honolua cinder cones on the horizon west of the Isthmus of Maui, near the island's southwest coast.

Haleakalā dominates the view to the east. Its younger and gently sloping flank abuts West Maui's much steeper and more eroded flank. Gentle slopes at the base of the steep flank are alluvial fans that spread from the mouths of its valleys and support vast fields of sugarcane and pineapples.

Hawai'i 30 reaches the southern shore of the Isthmus of Maui at Mā'alaea. Just south of milepost 7, a big outcrop exposes reddish black basalt rubble, which covers an ancient red laterite soil developed on older lava flows. Thicker layers of clinkery rubble separate thin flows of brownish black basalt toward the bottom of the outcrop.

Black Honolua basalt covers gray ash on top of a layer of red laterite about 2 feet thick, which has weathered in an older lava flow. Along the west side of Hawai'i 30 near Mā'alaea, about 0.3 miles south of milepost 8.

A lava flow of gray to red weathered trachyte at McGregor Point, the southernmost tip of West Maui.

McGregor Point

The pullout at McGregor Point, between mileposts 8 and 9, provides excellent views of Lānaʻi, nearly 20 miles to the west, and of Kahoʻolawe, 15 miles to the south. The tiny islet in the ʻAlalākeiki Channel between Kahoʻolawe and Maui is Molokini. It is the summit of a cinder cone that erupted on the submerged extension of Haleakalā's southwest rift zone during rejuvenated volcanism. The flooded crater is a marine preserve and popular snorkeling area. Other Hāna cinder cones stud the rift zone where on land. The largest of these, Puʻu Ōlaʻi, is at the coast. A 1790 lava flow from Haleakalā entered the ocean south of Puʻu Ōlaʻi.

Pale gray trachyte, dated at about 1.16 million years, is well exposed in outcrops and roadcuts near the McGregor Point pullout. It erupted during late-stage, Honolua, volcanic activity. The massive trachyte shows excellent flow banding, created as the magma sheared internally while it oozed down the slope. Small vesicles form zones, each several inches wide, that separate the denser layers. Seaward, the trachyte covers much darker basalt broken into an elegant palisade of vertical columns. It erupted during the main, Wailuku, stage of volcanic activity.

Watch for a roadcut about half a mile east of McGregor Point where a flow of late-stage alkalic basalt is visible. It covers a few feet of volcanic ash, weathered yellow. Heat from the flow baked the top layer of ash red. The ash covers a flow of Wailuku basalt.

Look east at milepost 10 for thin flows of Wailuku basalt, pāhoehoe full of lava tubes 2 to 5 feet across. These developed when the molten interior of the flow broke through a hard crust of cooling lava, then drained down the slope. Erosion and road construction have exposed the tubes.

Watch just west of the highway tunnel for a basalt flow that contains many large crystals of glassy green olivine, and a few of black pyroxene. Between the tunnel and Pāpalaua State Wayside, you may also see a flow of basalt filling a channel that eroded in older Wailuku basalt during the Honolua stage of volcanic activity.

Ukumehame Valley

Pullouts near milepost 13 provide a good view up Ukumehame Valley. The head of the valley is eroded into the caldera of West Maui, where the rocks contain a complex network of dikes associated with the volcano's main magma chamber. Ukumehame Stream carried abundant moonstones from this area to the coast. You can find a few in the beach sand, which consists mainly of black basalt pebbles and red cinder mixed with pieces of white coral.

From the highway, you cannot see the caldera's dense network of dikes, but you can see a satellite vent, Puʻu Koaʻe. It is a lava dome of trachyte nearly 600 feet high, part of the Honolua volcanic formation. Look in the western wall of the valley, a few miles away, to see it exposed in eroded cross

Ukumehame Valley.

Black basalt pebbles and red scoria are mixed with rounded coral fragments in the beach near the mouth of Ukumehame Valley.

section, along with the dike that fed lava into the dome. If you look closely with binoculars, you can see that the lava flows of older Wailuku basalts surrounding the dike are bent upward and broken, presumably by the force of intrusion. Other Wailuku flows exposed in the eastern side of the valley dip seaward.

Cones and Beaches Just South of Lahaina

Look northeast near milepost 15 for Puʻu Kīlea, a cinder cone that is part of the young Lahaina volcanic formation. Alkalic basalt that contains large olivine and small pyroxene crystals flowed out of the base of the cone. Hawaiians in ancient times carved petroglyphs on the surface of the cinder cone, and people in modern times have left graffiti.

A well at the mouth of the nearby valley produces 92-degree water. The heat probably comes from rocks still warm after the intrusion of Lahaina magma.

The upper part of the beach next to the highway near milepost 16 consists of dark gray pebbly basalt, red cinder, and some white coral. Black sand on the lower summer beach contains tiny grains of glassy, yellowish green olivine. Large crystals of clear olivine are cut into gemstones called peridot.

A gravel pit exposes the sloping internal layers of a Honolua cinder cone north of milepost 16. Farther north, inland from the highway, is Puʻu Mahānalua Nui, a dome of pale Honolua trachyte. The seaward flank split during the final phase of its extrusion, releasing a small flow of viscous lava. The flow is thick because trachyte lava is too viscous to run out into a thin sheet.

The beach near milepost 17 is mostly basalt pebbles and coral fragments. The beach at Launiupoko State Wayside is gray sand. Near shore, the bottom is mixed rock and sand.

Near Lahaina town, look up the slope to see Puʻu Laina, another cinder cone that grew during the greatest of the rejuvenated-stage eruptions. A much larger cone of older Honolua basalt rises farther inland. Both cones divert stream valleys, so the streams must have been flowing before the eruptions. Cinder blocked the stream channels, forcing them to flow around the cones. The broad, smooth slopes on the divides between the valleys are remnants of the original flank of West Mauiʼs volcanic shield.

Lahaina

Lahaina residents get their water from runoff, rain catchment, and wells. One well near town produces 82-degree water. Like the warm-water well near Olowalu, it probably is heated by rocks still warm from rejuvenated volcanism.

Offshore, a fringing coral reef extends about a mile north and south of the marina at Lahaina. It shelters the quiet water near shore, which is a good place for snorkeling and surfing.

Puʻunoa Point, at the north edge of Lahaina, exposes Lahaina lava flows. Some are ordinary alkalic basalts rich in little green crystals of olivine. Others are nephelinite, a rare kind of basalt so extremely rich in sodium that it contains a mineral called nepheline in place of the usual plagioclase feldspar. Unless viewed under a microscope, nephelinite looks like any other kind of basalt.

Hawaiʻi 30
Lahaina–Honokōhau Bay
9 miles

Almost 3 miles north of Lahaina, a bit south of milepost 24, a road leads to a resort at the base of Puʻu Kekaʻa, northernmost of the four Lahaina cinder cones. A walkway at the north end of the beach along the south side of Puʻu Kekaʻa provides a good view of the interior structure of the cone, which the waves have opened for inspection. You can see basalt bombs, blocks, spatter, and cinder.

The famous resort beaches Kaʻanapali and Kahana are expanses of calcareous sand eroded from the coral reefs offshore.

A mixture of spatter and cinder in Puʻu Kekaʻa, north of Lahaina.

The island of Molokai rises from the sea ten miles across Pailolo Channel. Though older than West Maui, the shield of east Molokai volcano still retains much of its original shape.

At Nāpili Point, basalt is exposed in a wave-cut bench, complete with tidal pools where the resident mussels, periwinkles, and sea urchins bore holes into the solid rock. A public access road leads to a small parking area at Nāpili Bay. There, a fine variety of tropical fish and coral flourish on old coral heads broken up during storms.

Fleming Beach

Near Kapalua, just northeast of milepost 30, shoreline outcrops next to Fleming Beach are lava flows from late-stage Honolua volcanic activity. They erupted between 800,000 and 500,000 years ago.

Roadcuts a mile or two north of Kapalua reveal weathered basalt and a red laterite paleosol, an ancient, buried layer of soil. The steep trail below a small parking area about 200 yards north of milepost 32 leads to another pleasant calcareous sand beach. Cliffs at the south end of the beach contain a bed of gray ash covering black gravel, which in turn caps another thin red paleosol, weathered on an older pale gray ash. All of these rocks are part of the Honolua volcanic formation.

Honokōhau Bay

Thick layers of gray trachyte ash and gray bouldery gravel are exposed in roadcuts within a half mile east of milepost 33. A small pulloff on the ocean side of the road at mile 35.6 provides a spectacular view northeast across Honokōhau Bay to red sea cliffs. The prominent thin, wavy layers exposed in the cliffs are in the interior of a large cinder cone that the waves opened to view. Roadcuts across the road expose basalt flows and volcanic ash.

A spectacular cut along the west side of the road near milepost 36 reveals a small dike cutting through a bed of volcanic ash and cinders. Much of the rock has altered to yellow palagonite, a clay that forms when molten basalt reacts with water and steam. The palagonitized volcanic ash and cinders settled on a rubbly surface, probably a wave-cut bench. The vent is probably in the bay.

The dike fills the fissure that the magma followed to feed the overlying lava flow. The red zone at the top of the volcanic ash layers may be laterite, but more likely it was baked red in the heat of the flow above.

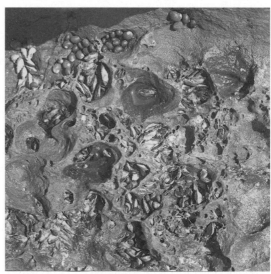

Mussells and snails in a rock shelf at Nāpili Point, near Kapalua.

Wave erosion opened the interior of this cinder cone along the northeast shore of Honokōhau Bay.

Look along the shore west of the parking area at Honokōhau Bay to see the same sequence of black basalt over yellow palagonite beautifully exposed in the cliffs above the modern wave-cut bench. These layers accumulated on the rubbly surface of an older wave-cut bench.

The West Maui Windward Coast
Hawai'i 340, The Kahekili Highway
Honokōhau–Kahalui
22 miles

Hawai'i 340 leads around the north cape of Maui below slopes dotted with Honolua cinder cones. The road between Honokōhau and Wailea Point is winding and very narrow, truly an adventure in motoring. When the road is not closed because of erosion, you can drive through to Wailea Point, where the challenge fades. Mileposts measure from zero at Kahalui.

As elsewhere in Hawai'i, the contrast between windward and leeward coasts is striking. Alluvial fans deposited along the dry leeward shore of West Maui make a gently sloping coastal lowland that extends offshore in a shallow bottom. Alluvial fans form mainly in dry areas, where stream flow is not sufficient to carry all the eroded sediment to the ocean.

The windward coast receives heavy rainfall, so streams here have enough flow to carry eroded sediment all the way to the ocean. That is one reason you see neither alluvial fans nor the gently sloping coastal lowland they form. Another reason is that the bottom offshore here drops steeply into deep water.

The trade winds and North Pacific gales build heavy waves that break directly onto the windward coast. The result is a cliffy shoreline that allows little easy access to the surf and no safe swimming places. The shoreline faces of Pu'u Koa'e and Mōke'ehia Island are fine examples of coastal erosion.

Big roadcuts on the east side of Honokōhau Bay, about a mile beyond the head of the bay, expose remnants of the bright red cinder cone that you can see by looking east across the bay. A little farther east, the layers become brownish gray ash. Because this ash was farther from the vent, it escaped the attention of the hot water, steam, and other gases that altered the rest of the cone.

A roadcut west of milepost 20, about half a mile east of the turnoff to the Nākālele Point Light Station, reveals basalt in the process of weathering into rounded residual stones. This flow lies on a red laterite weathered onto another basalt flow, which lies on a thin zone of red rock that it baked on yellow volcanic ash. The ash covers still another basalt flow. All this accumulated sometime between 1.5 million and 800,000 years ago, during the eruptions responsible for building the Wailuku shield.

Roadcuts along the 4 miles southeast of the Nākālele turnoff expose flows of Wailuku basalt. In places, you can see layers of yellow palagonite ash, in each case baked red on top in the heat from an overlying basalt flow. Elsewhere, the periods between lava flows were long enough for red soils to form.

Puʻu Koaʻe

A roadcut about a mile west of the tiny community of Kahakuloa exposes layers of gray volcanic ash that erupted during the late, Honolua, stage of volcanic activity.

Puʻu Koaʻe, the hill on the east side of Kahakuloa Bay, next to the village, is an eroded dome of pale trachyte from the Honolua stage of activity. The sea cliff exposes a well-defined section through the dome. It shows that the trachyte lava oozed into the crater of a broad pumice-and-cinder cone, which grew during an earlier stage of the eruption when highly concentrated gas and steam escaped. White trachyte ash from the explosive phase of the eruption overlies red laterite paleosol along the road about half a mile east of Kahakuloa.

A series of roadcuts between mileposts 12 and 4 expose pale gray trachyte ash, probably erupted from Puʻu Koaʻe. The more southern roadcuts are huge, and the ash in them has partially cemented into fairly solid rock. Intersecting fractures outline polygons in the rock, and rust-colored bands make crude bull's-eye patterns in them. The rusty color is iron oxide, which diffused into the ash from the fractures and then precipitated as concentric color rings called diffusion bands.

Diffusion bands in trachyte ash stand out in roadcuts between Kahakuloa and Waiheʻe.

145

A crossbedded dune rock near Waiheʻe.

Sand Dunes

A tract of sand dunes forms a prominent ridge along the ocean side of the road southeast of Waiheʻe, at milepost 4. They are made of pale sand much like the calcareous beach sand along the Lahaina coast. They consist of grains of broken coral and calcified algae that waves broke from offshore reefs and carried onto the shore. The dunes grew and migrated to their present positions during the most recent ice age, when sea level was lower and broad stretches of coral and sand were exposed. The rise in sea level at the end of the ice age submerged much of the original sand supply. Plants have stabilized large parts of the dune field.

Look northeast at the intersection with Hawaiʻi 330, just northwest of milepost 2, to see the roadcuts about 100 yards away. They expose sand dunes solidly cemented into limestone. The conspicuous thin layers that dip at various angles are crossbeds. The steepest layers of sand accumulated on the leeward faces of the dunes, where sand slides down the surface after blowing over the crest; the more gently inclined layers built up on the windward faces, where the wind drives the sand across the dune. A quarry at milepost 1 produces dune sand for construction aggregate.

Look east across the bay in the area near Waiheʻe to enjoy a splendid view of the northern flank of Haleakalā. You can see the broad slope of its shield arching gently down to the coast. The cinder cones of Kula basalt dotting the slope all erupted from the northwestern rift zone.

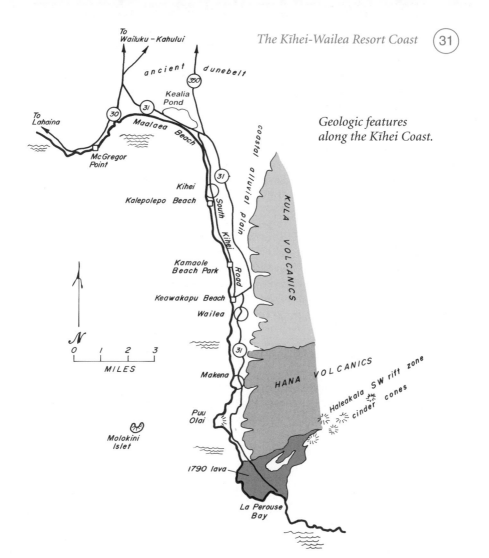

*Geologic features
along the Kīhei Coast.*

The Kīhei-Wailea Resort Coast
Hawai'i 31, Pi'ilani Highway
Hawai'i 30–Wailea
11 miles

Hawai'i 31 skirts East Maui's popular southwest resort coast. It provides access to South Kīhei Road and crosses the coastal section of Haleakalā's southwest rift zone, toward the site of Maui's most recent volcanic eruption. Lack of moisture and poor soil make the scrubby inland slopes unsuitable for farming, although much of it is cattle range.

The first two miles south from its junction with Hawai'i 30, Hawai'i 31 crosses weathered sand dunes on the Isthmus of Maui. Plant cover has stabilized the dunes. The sand came from reefs along West Maui's northeast shore.

The road reaches the coast at Māʻalaea Beach, a beautiful stretch of beige calcareous sand about 3 miles long and 75 feet wide. The inland edge of the beach has some outcrops of beach rock, and some small sand dunes. The bottom slopes steeply offshore, but conditions are usually fine for swimmers.

About a mile of road follows the top of a sand bar separating Keālia Pond from the ocean. When sea level rose to its present height at the end of the latest ice age, the sand bar did not exist, and Keālia Pond was a shallow bay. Since then, ocean waves have built up the bar across the mouth of the bay, isolating Keālia Pond from the ocean.

Just southeast of Keālia Pond, halfway between mileposts 3 and 4, Hawaiʻi 31 intersects with Hawaiʻi 350. From here, Hawaiʻi 31 swings inland for about half a mile, then continues parallel to the coast.

Outcrops near the northern end of the inland route expose basalt flows that have weathered to a rusty color. They are Kula basalts that erupted during the late stage of activity in Haleakalā Volcano, between about 700,000 and 350,000 years ago. More flows from the same period are exposed in roadcuts near mileposts 7 and 8. The flows issued from the sloping flank of Haleakalā's southwest rift zone.

Watch for big roadcuts in flows of black basalt near the southern end of the route, between mileposts 10 and 11. These are Honomanū lavas, which helped build the main shield of Haleakalā, between about 900,000 and 700,000 years ago.

South Kīhei, Alanui Road
Kīhei – Wailea – ʻĀhihi – Kīnaʻu Reserve
6 miles to Wailea

South Kīhei Road follows a series of beautiful beige sand beaches between small rocky points of basalt. A continuous row of tourist facilities lines the coast, but county parks preserve many of the beaches for public use.

This stretch of coastline has suffered severe erosion in recent decades. Waves have driven beaches as much as 300 feet inland over the past 40 years. Most erosion comes during fierce winter storms, called Kona storms because they blow in from the southwest, or Kona direction. Many efforts are under way to salvage shorelines and protect condominiums.

At milepost zero, the northern end of the route, South Kīhei Road passes condominiums and hotels facing a long beach of pulverized reef sand, Maipoinaʻoe Iaʻu Beach Park.

The rocky point that protrudes into the surf about 4 miles south of the Hawaiʻi 31 junction is basalt, part of a lava flow that erupted from Haleakalā's southwest rift zone during late-stage activity. Most of the basalt exposed in

Eroded remnant of a lava tube along the South Kīhei shore.

the area between Kīhei and Wailea also was erupted around this time, between about 700,000 and 350,000 years ago. The cooling lava trapped steam bubbles and other volcanic gases, forming the many vesicles.

Three sections of Kamaʻole Beach County Park between mileposts 4 and 5 separate South Kīhei Road from the ocean. The rocky points between Kamaʻole Beach Parks II and III are eroded remnants of black pāhoehoe lava containing many lava tubes.

A sort distance south of Wailea Shopping Village is Wailea Beach. Waves here have eroded rubble and cinder from beneath a thin basalt flow, undercutting it to make a scalloped beach with undertow runnels.

Scalloped Wailea Beach.

149

Puʻu Ōlaʻi.

Puʻu Ōlaʻi

A mile and a half south of the Shopping Village, watch for the first good views of Puʻu Ōlaʻi, a prominent cinder cone on the coast. Rocks exposed at the turnoff to Paipu Beach are older flows of ʻaʻā basalt covered with grass and trees. The turnoff to Mākena Landing leads to two small calcareous beaches with a fresh, black lava tube extending offshore.

A little farther south, the road passes the inland side of 350-foot Puʻu Ōlaʻi. It grew sometime within the past 100,000 years, during Haleakalā's rejuvenated, or Hāna, stage of volcanism.

The earliest stage of the Puʻu Ōlaʻi eruption was explosive. As rising magma reached the surface, expanding steam and gas blew the magma to shreds, which piled up and built the cone. After the vapors escaped, the remaining magma erupted quietly as a lava flow. You can follow a trail out to the point at the south side of Puʻu Ōlaʻi, where erosion has exposed the vent that produced the lava flow. It is a black basalt dike low on the flank of the cinder cone. Basalt fills the fissure that fed magma into the lava flow. You can still see the connection between the dike and the lava flow, the point where the magma erupted to become lava.

ʻĀhihi-Kīnaʻu Natural Area Reserve

The road enters the ʻĀhihi-Kīnaʻu Natural Area Reserve about 10 miles south of Wailea Shopping Village. It crosses a field of clinkery ʻaʻā lava and in some places skirts a beach of basalt and coral pebbles. The plant cover here is sparse, making it easy to trace the black river of basalt up the slope toward

Pu'u Ōla'i has a good example of a dike (left) *feeding into a lava flow, the layered rock at the top.*

its source. The flow is alkalic basalt of the Hāna series, full of pretty little grains of green olivine.

A little more than a mile inside the 'Āhihi-Kīna'u Reserve, look inland for a view up Haleakalā's southwest rift zone. The barren flow of chunky 'a'ā lava in the foreground erupted in 1790 from a split cinder cone low on the flank of the volcano. Farther into the reserve, the road crosses half a mile of an older 'a'ā flow covered with brush. It ends on the flow of 1790.

The crescent outline of Molokini Island 3 miles offshore is the eroded top of a cinder cone that was almost totally submerged when sea level rose at the

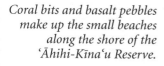

Coral bits and basalt pebbles make up the small beaches along the shore of the 'Āhihi-Kīna'u Reserve.

151

An 'a'ā basalt flow in the 'Āhihi-Kīna'u Reserve. Young cinder cones on Haleakalā's southwest rift zone are scattered across the mountainside in the background.

Cinder beds are visible in the eroded rim of Molokini, a Hāna cinder cone in the channel offshore from the 'Āhihi-Kīna'u Reserve.

end of the latest ice age. The low northern rim of the crater is visible as a pale streak of shallow water curving off the point at the west end of the island. The island is made of thin layers of dark brown to black cinders that fall away on both sides from the rim of the cone.

Molokini Island is a Marine Life Conservation District Seabird Sanctuary, and a popular destination for snorkeling and sightseeing boat trips from Maui. You can see many kinds of birds and fish, as well as the rocks. The underwater scene features coral heads in a rainbow of colors, along with a menagerie of colorful sea life, including yellow butterfly fish, white Moorish Idols, and dark gray and orange surgeon fish.

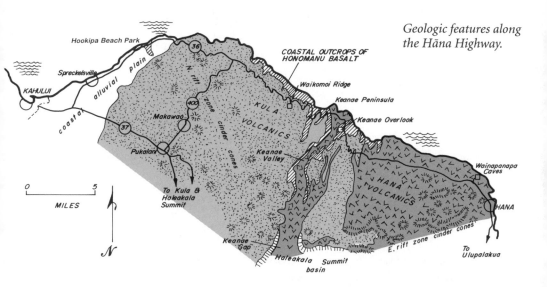

Geologic features along the Hāna Highway.

Hawai'i 36 and 360, Hāna Highway
Kahului – Hāna – Kīpahulu Valley
58 miles

You can drive all the way around Haleakalā in one long day, but the southern part of the route is too rough for some vehicles, and for many drivers. Most people drive the stretch of road along the northeast coast to the Kīpahulu section of Haleakalā National Park and back. This is the famous Hāna Highway, which winds along the wet windward coast, dipping in and out of deep channels eroded in the gently sloping flank of Haleakalā Volcano. For most of the distance, it passes through seacoast cloaked in tropical forest.

153

Inland from Hoʻokipa Beach and the Hāna highway, Kula cinder cones on the northwest rift zone of Haleakalā dot the skyline.

Heading east from Kahului, Hawaiʻi 36 passes through Spreckelsville, named for sugar magnate Claus Spreckels, who set up his first plantation on Maui in 1878. Large dunes of calcareous sand that eroded from offshore reefs flank the beach. The largest dunes are near the Maui Country Club, a few miles east of Spreckelsville. Early Hawaiians used the dunes as a burying ground.

In January 1938, about 25 miles north of Spreckelsville and 65 miles below the surface, the earth generated what the Hawaiʻi Volcano Observatory classed as a moderate to strong earthquake, the strongest to strike Maui in recorded history. The shock jiggled seismographs as far away as New York. Damage was slight and no tsunami followed. The cause is a mystery; perhaps it relates to the sinking of the island. Though it is well north of the Hawaiian hot spot, Maui is settling into the ocean floor at a rate of about an inch every decade.

H. A. Baldwin Park hosts a popular beach on the windward side of Maui. The long strand of beige reef sand here is one of the best places on the island for bodysurfing and beachcombing. Outcrops of beach rock are scattered along the shore, especially toward the west end.

Along the 7 miles east of Kahalui, the road crosses the alluvial plain of the Isthmus of Maui. Sugarcane planted on reddish brown laterite soil covers the gentle slopes. Watch for prominent cinder cones standing along the sloping skyline to the east, on the northwest rift zone of Haleakalā.

A few miles east of Lower Pāʻia, the road skirts the top of sea cliffs overlooking rocky points between sandy beaches. Farther east, it winds across hills in the northwest rift zone.

Mudflow deposit at Hoʻokipa Beach Park.

Near Hoʻokipa Beach Park, a world famous windsurfing mecca, the road crosses a small area of alkalic basalt flows west of milepost 9. It erupted from the northwest rift zone sometime between 700,000 and 350,000 years ago. The young cinder cones farther east and inland erupted during the Kula stage of volcanic activity. This least active of Haleakalā's three rift zones erupted very little during the rejuvenated stage of volcanism.

Hoʻokipa Beach, like most along this shore, is made of eroded reef sand. A strong longshore current keeps the sand moving. Outcrops near the beach expose a mudflow deposit of rounded boulders of vesicular basalt embedded in hardened ash and mudstone. The mudflow that carried this mess came out of an interior valley. Mudflows can transport big boulders like these because mud is much denser than clear water and exerts a correspondingly greater buoyant effect. The prominent pockmarks in these rocks form by salt weathering. Salt spray soaks into the rock and dries up. As tiny salt crystals form, expansion pries off a piece of the rock, leaving a depression. Moisture collects in the depression and gradually enlarges the hole.

Just east of Hoʻokipa, the road heads inland a mile or two, passing through the upland pineapple fields.

Hawaiʻi 36 meets Hawaiʻi 365, the Makawao Road, east of milepost 16, where it degenerates into Hawaiʻi 360. The mileposts start again from zero. Just east of the intersection, the road passes through some big cuts in brown volcanic rubble, with a red laterite paleosol about a foot thick sandwiched in the middle. The lush tropical growth and deep erosion on this exotic coastline contrast vividly with Haleakalā's dry southwest shore.

Hawaiʻi 360 crosses countless gulches cut into the slope of the volcanic shield. The exposed rocks are mostly in the Kula volcanic formation. Older

Keʻanae Point from Kaumahina State Wayside.

Honomanū volcanic rocks that erupted in the shield-building stage of activity lie at the gulch bottoms and in the lower walls of many gulches, and at the base of the sea cliffs.

Just east of milepost 6, erosion has worn a massive Kula basalt flow into boulders in the stream bottom at an old diversion dam. Look downstream from the highway bridge to see vertical potholes about a foot in diameter. Rocks swirling on the bottom beneath persistent eddies in the stream carved these potholes out of the basalt bedrock.

Near mile 11, the road crosses Puohokamoa Stream. A trail to the left of the first pool leads to a larger pool and Puohokamoa Falls. It pours over a resistant ledge in the massive core of an ʻaʻā basalt flow.

Kaumahina State Wayside is near mile 12. Look east along the coast from the upper part of Waiakamoi Ridge Nature Trail for a view of the Keʻanae Peninsula.

The road loops east around the high cliffs that embrace Honomanū Bay. Near milepost 15, the highway skirts the steep, rugged coastline, passing roadcuts in black Kula basalt. If you watch carefully, you can spot a few pāhoehoe surfaces between patches of moss.

Keʻanae Valley

Just east of milepost 16, the road descends into the Keʻanae Valley. Beneath the lushly vegetated rims lie old Kula lavas. The middle valley walls are

Surf pounds the 10,000-year-old Hāna basalt at Keʻanae Point.

The northeast shore of Maui, near the Keʻanae Arboretum.

157

even older Honomanū basalt, the lavas that built the main shield of Haleakalā. Honomanū flows also make up the floor of Keʻanae Valley but have been mostly buried by younger sediments and Hana lavas originating from Haleakalā's summit basin, 10 miles to the south and 8,000 feet higher. On at least four occasions, very fluid Hana flows followed Keʻanae Valley to the coast, in places leaving only thin veneers of basalt plastered against the valley walls. The youngest flow, 10,000 years ago, built most of the modern Keʻanae Peninsula.

Keʻanae Arboretum, on the valley floor, contains beautiful specimens of tropical plants, and a kalo (taro) patch cultivated in the traditional Hawaiian style. The arboretum stretches along the bank of Piʻinaʻau Stream, which here and there has cut through the young Hana flows to expose patches of Honomanū basalt.

Waterfalls and Overlooks

Wailua wayside lookout near milepost 19 provides good views of Wailua Canyon, eroded deeply into flows of Kula lava. The Keʻanae Overlook, also near milepost 19, offers views of the upper Keʻanae Valley to the south, the ocean to the north, and fields of kalo below.

Waikani Falls is at mile 19.5, and a smaller falls is at mile 21. Near milepost 22, the highway reaches Kopiliula Falls, and a late-nineteenth-century sugarcane irrigation ditch that is also used as a domestic water supply. This ditch, and others that connect to it, begins about 2 miles to the east, continues west along the side of Haleakalā, then south most of the way to dry Kīhei, a distance of about 40 miles.

Waterfalls at Puaʻakaʻa State Park are held up by resistant basalt flows.

Two waterfalls drop across successive basalt flows into beautiful plunge pools in Puaʻakaʻa State Park. Watch for another waterfall on Hanawī Stream, near milepost 24.

Haleakalā East Rift Zone, and a Lava Tube

The landscape changes near milepost 24, where the road passes Haleakalā's east rift zone. Dozens of cinder cones and lava flows of Hāna and Kula basalt create a terrain hundreds of thousands of years younger than the one on the older volcanic rocks.

The hummocky countryside between mileposts 27 and 28 is typical of the topography that develops on hawaiite, one of the kinds of alkalic basalt that generally erupts during late-stage volcanic activity.

Side Trip to Waiʻānapanapa State Park

The road to Waiʻānapanapa State Park turns off the Hāna Highway at milepost 32, about half a mile from the Hāna airport. You can explore a large lava tube at Waiʻānapanapa; parts of the roof have collapsed and formed open skylights. Pools of beautifully clear water inside are popular swimming holes. They contain brackish groundwater, which at this low elevation lies close to the surface.

A small black sand beach is below the parking area at Waiʻānapanapa. Unlike the black sand beaches that form where molten lava explosively enters the sea, the sand here comes from wave erosion of basalt in the sea cliffs. This continuous process will probably keep the beach supplied with black sand for

Sea arch at Waiʻānapanapa.

a long time. Look a few hundred yards southeast along the coast to see an arch that the waves have eroded in the basalt.

At mile 33.8, the highway meets an alternate coastal road to Hāna Bay and its pretty beach—700 feet of mixed lava and coral sand between points of Hāna basalt. This is one of the safest swimming beaches in East Maui. Snorkeling is good close to the shore, between the pier and lighthouse.

Look upslope to the west at mile 34.2 to see Puʻu o Kahaula, an older cinder cone of Kula basalt sticking up through younger flows and cinder that buried the lower flanks. A side street leads to a lookout on top. It is at the crest of the east rift zone, one of the longest rift zones in the Hawaiian Islands. The volcanic ridgeline extends east nearly a hundred miles from the summit of Haleakalā to the bottom of the Hawaiian Deep.

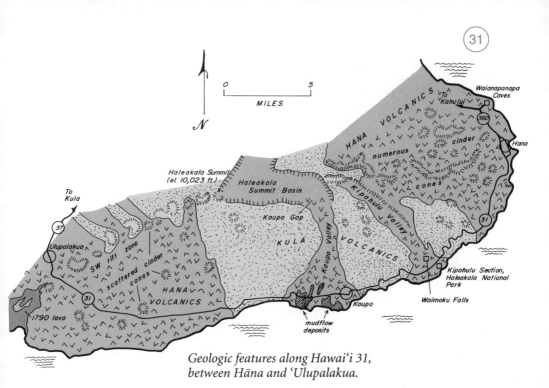

Geologic features along Hawai'i 31,
between Hāna and 'Ulupalakua.

Hawai'i 31
Hāna–Haleakalā National Park, Kīpahulu Section
9 miles

Beyond Hāna, toward Kaupō, the mileposts count down from 52.

At milepost 51, look upslope a half mile or so to see Pu'u Kolo, an old cinder cone of Honomanū basalt. At mile 50.4, the road skirts the inland flank of still another cinder cone, one where the sea flooded the crater. A popular beach lies along the western wall of the crater. 'Ālau Island, a seabird sanctuary offshore, is another eroded vent.

Between Hāna and Kīpahulu

Stepping out to explore the rolling pasture land near milepost 48, you may discover some ancient petroglyphs etched in lava. At mile 46.3, a roadcut exposes red laterite beneath a flow of 'a'ā basalt.

South of mile 46, the highway enters older terrain eroded on Kula basalts. At mile 44.7, the highway passes Wailua Falls. The stream drops 95 feet over a palisade of vertical basalt columns. Near mile 43.5, roadcuts in pāhoehoe basalt expose sections of several small lava tubes.

South of milepost 43, the road enters Haleakalā National Park. Within a mile, it reaches Palikea Stream at the mouth of Kīpahulu Valley, one of the wildest and most remote valleys in Hawai'i. Flows of Hāna lava poured from the summit of Haleakalā down Kīpahulu Valley to the ocean. They lie beneath

*Columns in a flow
of Kula basalt near
Wailua Falls.*

*Pools in the stream at the
mouth of Kipahulu Valley.*

most of the gently sloping valley floor. You can see these flows exposed in Palikea Stream Gulch.

Coastal trails in Kīpahulu provide access to the shore, to archaeological sites, and to Palikea Stream, which has many small waterfalls and large plunge pools. The waterfalls spill over ledges, each one the massive core of an 'a'ā basalt flow. The waterfalls eroded the plunge pools in the rubble at the base of each flow. Look around the biggest pool coastward from the road for a red laterite paleosol sandwiched between lava flows. Many years passed to weather this ancient soil on the flow beneath it before the flow above buried it.

A 2-mile trail leads inland to a young valley with a broad amphitheater at its head. Pipiwai Stream eroded it into Hāna and older Kula lavas. The trail passes through spectacular forest, stream, and waterfall scenery.

Stream spills over basalt flow ledges, Kīpahulu Section, Haleakalā National Park.

Hawai'i 31, South Flank of Haleakalā
West of Kīpahulu – 'Ulupalakua Ranch
27 miles

West of Kīpahulu, the road deteriorates into a track unsuitable for general travel. The countryside becomes drier and less hospitable in the rain shadow of Haleakalā. At the mouth of Kaupō Valley, a magnificent view opens inland up steep slopes into the eastern part of Haleakalā's summit basin, where Hāna basalts nearly filled the broad head of the valley.

The steep mountainside, abundant loose volcanic ash and cinder, and a climate that produces occasional heavy rains provide ideal conditions for mudflows in Kaupō Valley. Rain mixes with volcanic ash and yields dense mud, easily capable of moving large boulders. Hāna lava flows and mudflows fill most of the valley floor all the way to the coast, where they form a wide shelf. One prehistoric mudflow is nearly 300 feet thick near the lower end of the valley.

Because the flanks of Haleakalā are much steeper along this coast than elsewhere on Maui, the summit is closer to the coast here than to the coast in other parts of the island. The volcano rises from the ocean like a fortress. Some people feel uncomfortable here, trapped between the enormously strange landscape to the north and the vast ocean to the south. If the air is clear, you can see the landmass of the Big Island across the channel, adding to the immensity of the surroundings.

Lo'alo'a Heiau is an ancient temple and a National Historical Landmark. The road west of it crosses a series of deep gulches eroded through thick stacks of Kula 'a'ā. Herds of feral goats overgraze the slopes, exposing the red laterite soil to extremely rapid erosion from rain splash and surface runoff.

About 7 miles west of Kaupō, the road begins climbing toward the southwest rift zone, entering less arid country and crossing young Hāna lava flows that erupted from the rift zone on the slope far above. Gravel quarries and a roadcut near milepost 22 expose the innards of a big red and rusty brown cinder cone, one of the Luala'ilua Hills. Layers of black and rusty brown cinders exposed in the eastern quarry record the flank of the cone as it was during the various stages of development, as hot cinders blown out of the vent slid and tumbled down the slopes.

Between mileposts 21 and 20, the road passes half a mile downslope from Hōkūkano, another cinder cone surrounded by a younger flow of Hāna basalt. Look upslope to the north or northwest to see a line of cinder cones marking the southwest rift zone, the source of the flows. About half a mile downslope, near milepost 18, you can see Pīmoe, another prominent cinder cone.

Near mile 16, the road reaches the crest of the southwest rift zone, the main source, together with the east rift zone, of Haleakalā's young Hāna lavas.

The cinder cone half a mile up the slope from the road is Puʻu Māhoe, elevation 2,660 feet. It erupted in 1790. Black lava flows issued from the base of the cone and flowed all the way to the ocean.

About a mile down the slope from the road is Puʻu Naio, a nicely preserved, grassy cinder cone with a crater. Moisture condensing in the air rising from the coast falls as rain that supports the open woodland on the older soils. Hāna lava flows and cinder cones look especially fresh in the drier land below.

Kahoʻolawe

Kahoʻolawe lies about 10 miles to the southwest offshore from the southwestern cape of Maui. The island is 11 miles long and 6.5 miles wide. It is built mainly of tholeiitic shield basalt, with a thin cap and caldera fill of alkalic lavas that erupted during late-stage volcanic activity around a million years ago. They erupted from a rift zone that trends southwest from a caldera about 4 miles wide at the eastern end of the island.

Like all the other Hawaiian Islands, Kahoʻolawe has shed big pieces of itself in the form of huge submarine slides. A large part of the island that once extended from the caldera toward Maui is gone. Lavas erupted from Haleakalā or the Big Island cover the slide debris on the ocean floor. Eruptions from the rejuvenated stage have left small patches of lava in the caldera.

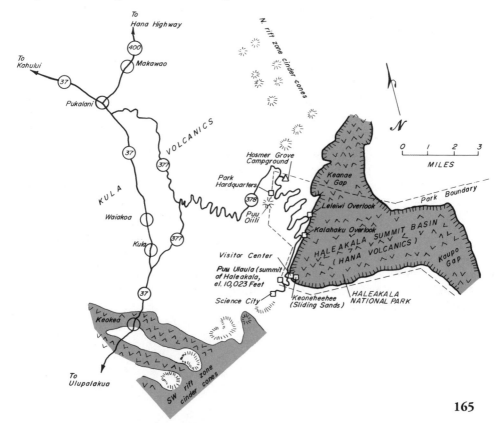

Kahoʻolawe rises about 1,500 feet above sea level. The upper slopes are covered with a thick mantle of red laterite soil. The soil is not to be found below about 800 feet. It probably was washed away by a giant tsunami in the wake of one of the cataclysmic landslides. The ʻAlikā landslide that dropped into the ocean from the west coast of the Big Island about 100,000 years ago is the likely culprit. What soil remains is continually blowing and eroding away because overgrazing and military bombing practice have destroyed the plant cover. Local geologists call the center of the island the Dust Bowl. Efforts to restore a drought-resistant plant cover are under way.

North of ʻUlupalakua Ranch, the road continues back to Kahului as Hawaiʻi 37, crossing the pastoral upland of the lower western flank of Haleakalā. This is one of the pleasantest landscapes in the Hawaiian Islands.

Hawaiʻi 37
Kahului–ʻUlupalakua, via Pukalani and Kēōkea
7 miles

Between Kahului and Pukalani, open fields along Hawaiʻi 37 provide intermittent views east to several pronounced knobs on the skyline, protruding from the smooth lower flank of Haleakalā. These are old, eroded Kula cinder cones along the northwest rift zone of the volcano.

Beyond Pukalani, the highway passes farms and ranches. Several roadcuts around mile 6.5, at the junction of Hawaiʻi 377, expose weathered basalt flows covered with a thick, reddish brown soil. The grassy roadcuts farther south show massive quantities of basalt and brown ash.

Look upslope near Kēōkea to see cinder cones at the crest of the southwest rift zone of Haleakalā. Although many of the cinder cones in this rift zone erupted during the Hāna stage of volcanism, these are part of the Kula stage of volcanic activity.

South of Kēōkea, Hawaiʻi 37 continues through the verdant Up-country, crossing eroded Kula basalts most of the way. This flank of Haleakalā slopes more gently than the other sides, because lava from it piled up against West Maui, building the Isthmus of Maui. A dry climate with low rates of stream erosion also helps explain why this landscape is so gentle. The basic shape of the land has changed little during the past 350,000 years.

Hawaiʻi 37 shrivels into Hawaiʻi 31 at ʻUlupalakua Ranch, the site of the Makee Sugar Mill, built in 1878. Just south of the ranch, one of the few wineries in Hawaiʻi makes wine from local pineapple juice and imported grape juice.

Hawai'i 377 and 378
Pukalani – Summit of Haleakalā
27 miles one-way to summit

Many thousands of visitors drive this winding highway to the rim of Haleakalā's summit basin every year. The road climbs the western flank of the volcano into ever steeper and drier country, then enters a subalpine environment. It finally reaches the summit in the southwest rift zone, over 10,000 feet above sea level.

Outside Pukalani, Hawai'i 377 heads up the mountainside from Hawai'i 37. Pu'u Pane, 3 or 4 miles beyond the junction, is a cinder cone that erupted during the Kula stage of activity. A quarry works it for road gravel. Just beyond Pu'u Pane, a flow of Kula basalt covers an old soil.

Mileposts on Hawai'i 378, the route to the Haleakalā summit, begin at zero at the junction. All the rocks along the road are Kula basalts that erupted from Haleakalā between 700,000 and 350,000 years ago. Most are alkalic basalt, rich in sodium and generously speckled with glassy grains of green olivine. Watch the roadcuts between 4 and 5 miles from the intersection for thin flows of vesicular lava. These slope parallel to the mountainside, showing that little has been lost to erosion since they erupted. Thin lava flows are characteristic of steep slopes.

Layers of ash and cinder on the flank of Pu'u Nianiau are exposed in a roadcut near the boundary of Haleakalā National Park.

A volcanic bomb embedded in cinder near Pu'u Nianiau.

Watch near milepost 7 for roadcuts where you can see Kula basalt flows interbedded with volcanic ash erupted from Pu'u Pahū, a nearby cinder cone. Another cinder cone of Kula age, Pu'u 'Ō'ili, erupted the ash that covers 'a'ā lava flows near mile 8.7.

A large gravel quarry at mile 9.5 has opened the interior of another cinder cone, Pu'u Nianiau. You can see the layers of volcanic cinders that blew out of the vent and settled on the flanks of the growing cone. Look for the larger bombs, and the sags in the layers below them. Younger ash covers an old erosion surface on one side of the quarry wall.

A big roadcut at milepost 10, just outside the boundary of Haleakalā National Park, shows a neatly layered deposit of volcanic ash that has weathered to a rusty color. It contains blobs of black basalt ranging in diameter from about 1 inch to 2 feet. They are full of vesicles that formed as the solidifying lava trapped gas bubbles.

Haleakalā National Park

The road enters the northwest rift zone of the volcano as it crosses the boundary of Haleakalā National Park. On a clear day, you can see cinder cones along the line of the rift zone all the way to the coast, 9 miles east of Kahului. Park Headquarters offers excellent exhibits and information about hiking trails in and around the volcano. The road winds for another 9 miles to the rim of Haleakalā's breathtaking summit basin.

At mile 12.5, the road switchbacks through a lava flow that has cracked into vertical columns about 6 inches across. At mile 13.5, the road again passes Pu'u 'Ō'ili, the cinder cone below the road. Between miles 15.3 and 16.6, the highway cuts through a vesicular basalt flow about 4 feet thick that lies on red laterite and cinders.

An ʻaʻā flow overlies red laterite and a cinder bed at mile 16.6 on the road to Haleakalā's summit.

Leleiwi Overlook

From Leleiwi Overlook, you have a good view of Haleakalā's summit basin, which is 7.5 miles long and 2.5 miles wide. Look oceanward from the parking lot to see West Maui, with the island of Molokaʻi beyond. The island of Lānaʻi is a bit south of west, and Kahoʻolawe is in the southwest. Tiny Molokini punctuates the strait between Maui and Kahoʻolawe.

Look about 100 feet down the road from the overlook for an outcrop of Kula basalt. The upper several feet of the flow is full of glassy green olivine crystals about the size of peas and black pyroxene crystals about twice that size. This mineral composition is typical of ankaramite, a rare variety of alkalic basalt. The lower part of the flow contains few large crystals and looks more like ordinary alkalic basalt. Both upper and lower parts appear to be separate lobes of the same flow. What accounts for the striking change in composition? Geologist Gordon Macdonald suggested that the eruption tapped first the upper part of the magma chamber, then the lower part, where the large crystals had settled. So now we see the lower part of the magma chamber in the upper part of the flow.

Kalahaku Overlook

You can see the tilted slabs from a broken lava flow around the Kalahaku Overlook parking area. They are in pressure ridges that developed when part of the flow slowed, causing the advancing lava from behind to buckle. The medium gray rock is alkalic basalt, part of the Kula volcanic formation.

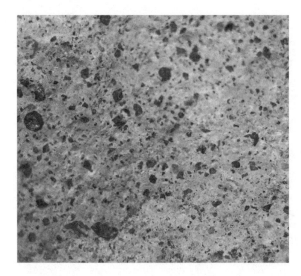

Large, black pyroxene crystals and smaller olivine crystals are embedded in ankaramite near Leleiwi Overlook.

Each of the successive views into Haleakalā's summit basin is different; the panoramas become more breathtaking with increasing elevation. Look northeast from the overlook to see the large gap Keʻanae Stream cut as it eroded the western half of the basin. Look east to see Kaupō Gap, 7 miles away, which Kaupō Stream eroded through the eastern half of the basin. Long after these streams eroded deep valleys, rejuvenated volcanic activity filled the valley floors with lava and cinder to make the present basin floor.

The rough alignment of cinder cones across the basin continues the row of Hāna vents from the southwest rift zone across the summit into the east rift zone. The cinder deposits owe their spectacular red colors to alteration by steam and volcanic gases during and shortly after the eruptions. Kula lavas make up the enclosing green walls.

No one knows exactly when rejuvenated eruptions began in the summit basin. A radiocarbon age date on charcoal from a buried soil bed suggests that as many as twenty eruptions have punctuated the past 2,500 years.

The basin floor lies 2,000 to 2,500 feet below Kalahaku Overlook. The highest of a dozen cinder cones rises about 600 feet. But these statistics do not convey an accurate sense of scale as vividly as an occasional glimpse of hikers on the trails far below.

In the distance to the south are the towering summits of Mauna Kea and Mauna Loa on the Big Island. Before Haleakalā sank deep into the earth's mantle, it was probably as high as they are, more than 13,500 feet.

Visitor Center and the Sliding Sands

The Visitor Center, near the rim of the volcano summit, offers an encompassing view of the basin. The road beyond ends in Puʻu Ulaʻula crater, the summit cone of Haleakalā, where huge lava bombs lie scattered on the slopes.

The vast erosional basin at Haleakalā's summit is partly filled with Hāna volcanic rocks, which create an eerie, multicolored landscape. This view is from Kalahaku Overlook toward the Sliding Sands at the western end of the basin.

Haleakalā's summit basin spreads out below the Visitor Center, located on the western rim.

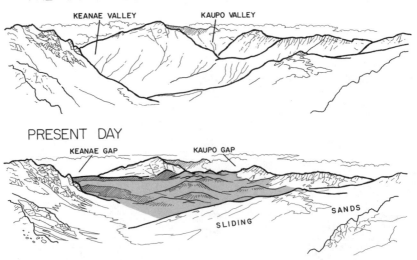

PRE-HANA VOLCANICS

KEANAE VALLEY   KAUPO VALLEY

PRESENT DAY

KEANAE GAP   KAUPO GAP

SANDS

SLIDING

Haleakalā's summit basin, viewed from the Visitor Center, before and after filling with Hāna volcanic rocks. The upper drawing is partly conjectural.

Escaping steam and gases power the explosive eruptions of cinder cones. The cones expand as magma rises to the surface and then blows from the vent, coughing up scraps of molten lava that cool as they fall through the air. The larger scraps fall close around a vent and build up a cinder cone; the smaller shreds drift on the wind as a cloud of ash. The last few explosions leave a summit crater shaped like a cup.

Imagine a cinder cone eruption from this safe perspective on the rim: Clouds of ash rapidly churn from a steaming black cone only a few miles away, rising like a dark tower into the sky above the basin floor. Incandescent bombs arc through the air from the ash clouds and fall as glowing globs of molten lava, tumbling down the slopes of the growing cone. An echoing roar reverberates from mountainside to mountainside. The ground shudders. Quite a scene for a romantically moonlit night.

From the summit, you can peer down the rugged spine of the southwest rift zone to the ocean, 15 miles away. Kahoʻolawe stands across the channel, together with Molokini, a largely submerged cone on the underwater continuation of the rift zone.

Science City, a research complex, perches on a nearby cone downslope from the Visitor Center. The first precise instrumental measurements showing that the Pacific Plate moves northwest at about 4 inches a year, carrying the Hawaiian Islands along for the ride, were taken here. Science City is also home to lunar and solar observatories, a satellite tracking station, and a radio repeater station.

The only way to see the interior scenery of Haleakalā's summit basin is on foot, most conveniently on the Sliding Sands Trail, which descends from near the visitor center to the basin floor. However, the following description takes you only as far as Kaluʻuokaōʻō, a young Hāna cone along the way. The vertical descent is 1,300 feet; the round-trip distance is 5 miles, at an elevation between 8,500 and 10,000 feet. Morning is best for this outing, before clouds fill the crater.

Keoneheʻeheʻe, the Sliding Sands, is a long cinder slope below a cinder cone of Kula basalt near the Haleakalā summit. Sliding Sands Trail starts across the eroded edges of several flows of Kula basalt with ʻaʻā surfaces at the basin rim, then switchbacks down the cinder ramp.

The volcanic landscape of barren, oxidized cinder seems as surrealistic as would the surface of another world. Volcanic bombs as much as 2 feet across litter the terrain; escaping gases blew them out of the vent as blobs of pasty magma full of steam and gas. Most of them are rounded, even streamlined. The broken ones show dense rims around highly vesicular cores. The dense outer rind chilled before the expanding gases inside could create vesicles like those in the cores.

Watch about halfway to the basin floor for the short side trail leading north to the rim of Kaluʻuokaōʻō, about a third of a mile away. It crosses a slope of volcanic ash that supports scattered ʻāhinahina plants, or silversword. This endangered plant grows only at the summit of Haleakalā. A related and even rarer species survives on the middle slopes of Mauna Loa on the Big Island.

The side trail ends in a partial circuit around the rim of Kaluʻuokaōʻō Crater, a cinder cone with slopes of fine fragments and scattered blocks and bombs. The prominent cone one-third of a mile due east is Kamoaliʻi. The large flow of black lava north of Kamoaliʻi was erupted from the break in its base. Debris from the nearby cinder cones does not litter the slopes, so it must be younger than they are.

Sliding Sands Trail winds past Keoneheʻeheʻe, the cinder-flanked crater in the lower left, on the way down to the floor of the summit basin inside Haleakalā.

173

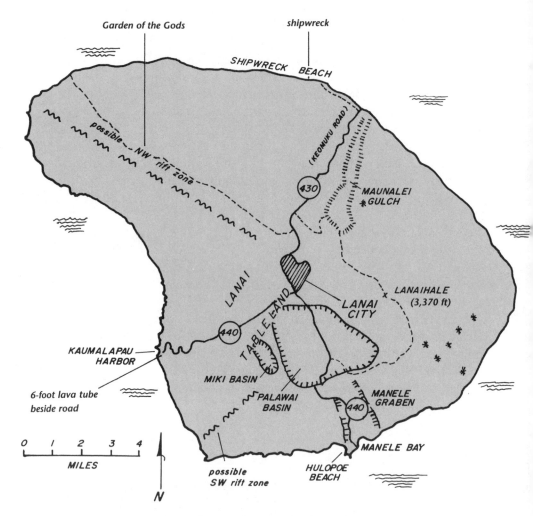

Major geologic and geographic features of Lāna'i.

4
Lāna'i, The Pineapple Isle

Lāna'i, the smallest of the easily visited Hawaiian Islands, is 18 miles long and 13 miles wide, encompassing only about 141 square miles. The entire island nestles in the dry rain shadow of Haleakalā on Maui, leaving it without a wet windward side. Jim Dole, who bought the island in 1921, plowed under the cactus growing in the Pālāwai basin, south of Lāna'i City, and planted pineapple. Pineapples are still grown on much of Lāna'i, but foreign competition has so intensified that many of the old pineapple fields are again open rangeland.

Lāna'i is a shield volcano that seems to have died as the main shield-building stage ended, without ever erupting any late alkalic basalts. The island is mainly a stack of tholeiitic basalt lava flows, and some of them are full of olivine crystals.

Three rift zones appear to radiate from the south-central part of the island. The northwest rift zone is the longest and was probably the most active; it built the island's prominent northwestern bulge. Several dikes appear in the walls of shoreline cliffs at the northwestern cape, where the rift zone meets the sea. Inland, only a few vents have survived erosion.

The southwest rift zone enters the sea at the island's southwestern cape. One pyroclastic cone still rises in the rift zone. You can also see an eroded lava shield and many dikes exposed in the sea cliffs.

The third rift zone extends southeast, where many small cones survive. A network of faults trending generally southeast along the axis of the rift zone border a sunken strip of land, the Mānele graben. Many of its dikes are exposed in the shoreline cliff at Mānele Bay.

Pālāwai and Miki Basins

Most Hawaiian geologists long identified the Pālāwai Basin south of Lāna'i City as the old summit caldera of Lāna'i. It is about 2.5 miles across and 100 to 150 feet deep. And they interpreted Miki Basin, along the western edge of the Pālāwai Basin, as the largely buried and eroded remains of a pit crater. The stack of thick, horizontal lava flows that form a bench about a mile wide east of Pālāwai look like the kind that erupt into a caldera. If so, the basin may simply have been a centralized subsidence area inside a larger depression,

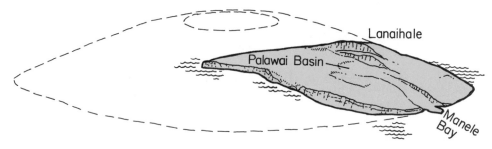

Lānaʻi is the remnant of a volcano that slid into the ocean.

like the central floor of Kīlauea caldera that collapsed during the nineteenth century.

Faults that follow the northwest rift zone and continue south to the Mānele graben seem to split Pālāwai Basin. About a third of the island, including the area where Lānaʻi City and the airport are, dropped hundreds of feet on the southwest side of these faults. This leads some geologists to look at Lānaʻi in a different light.

Some geologists suggest that a landslide completely removed the former summit of Lānaʻi and that the present island is the northeastern flank of a once much larger volcano. If so, the thick flows under the bench east of Pālāwai Basin could well be lava that ponded in depressions at the head of the slowly developing slide. Continued movement then broke through the flows, leaving the remnant that we see today. Pālāwai and Miki basins may be swales near the top of the giant slide. The pattern of rift zones may not be as simple as it appears at first glance.

Current evidence does not definitively establish or discredit either geologic interpretation of Lānaʻi. Nevertheless, we find the second view intriguing and join those who think that an enormous slide dumped much of the original island into the ocean.

In fact, startling evidence has recently been discovered on the neighboring ocean floor. Sonar maps of the ocean floor reveal a large rock mass stretching nearly 70 miles from the southwest coast of Lānaʻi, spreading into two huge lobes where it enters the Hawaiian Deep. Geologists call this the Clark Slide. It must have moved very fast to have traveled so far.

Hawai'i 440, Kaumalapau Highway
Lāna'i City–Kaumalapau Harbor
6 miles

The 6 miles of Kaumalapau Highway west of Lāna'i City cross the broad agricultural tableland of central Lāna'i, north of Pālāwai and Miki basins. Excellent views of Pālāwai Basin appear south of the highway a mile or so west of Lāna'i City.

The steep face of the bench along the eastern edge of Pālāwai Basin is about 450 feet high. Some geologists think it is one of the headwall scarps of the Clark Slide. If so, a big piece of the western side of Lāna'i dropped at least 450 feet, but not into the ocean.

About 5 miles west of Lāna'i City, Hawai'i 440 reaches the rim of the island's interior tableland. Roadcuts along the steep western flank of Lāna'i show the full thickness of the red laterite soil typical of the tableland, all the way down to bedrock.

Here and there, pinnacles of resistant bedrock project up into the soil, which ranges from a few feet to more than 10 feet thick. These buried rock pinnacles resemble the ones exposed by erosion along Keōmuku Road, Hawai'i 430, north of Lāna'i City. The soil is full of residual stones.

Near its western end, Hawai'i 440 winds down a steep slope to the coast, passing a series of roadcuts that expose both thin pāhoehoe and thicker 'a'ā lava flows. Most of these lavas erupted in the southwest rift zone, a mile or

A lava tube about 18 inches wide, in pāhoehoe basalt, is located just above milepost zero, east of Kaumalapau Harbor.

177

The west coast of Lānaʻi, north of Kaumalapau Harbor.

so south of the road. Ancient red laterite soils sandwiched between some flows near the coast show that long intervals of weathering separated eruptions during the last stages of shield growth on Lānaʻi.

Look a few feet above the south side of the highway at milepost zero for a prominent lava tube about 6 feet across, with the roof partly collapsed. Thin flows of black pāhoehoe basalt across the highway contain several small lava tubes 1 to 2 feet across. Lava tubes like these form when a crust hardens on top of a tongue of lava flowing down a gentle slope. Molten lava flows out from under the crust, leaving behind a hollow tube.

The road ends at Kaumalapau Harbor, offering a spectacular view along the precipitous west coast of Lānaʻi. The impressive cliffs continue around the southern coast of the island. The cliff probably began as a headwall scarp from the Clark Slide. Waves have been modifying it ever since.

Waves attack the cliff by first eroding a notch at sea level, primarily during the heavy winter storm season. The notch undermines the cliff face, and eventually it collapses, forming a new face a short distance inland. The waves then work the rubble seaward across the wave-cut bench at the foot of the cliff.

No cliffs of any origin face the narrow Molokaʻi and Maui channels along the northern and eastern coastline of Lānaʻi. That side of the island has not experienced large landslides nor is it exposed to heavy surf.

Just south of Kaumalapau Harbor lies the mouth of Kaumalapau Gulch, a small valley enclosed between steep walls. Kaumalapau Stream eroded this valley when the island stood substantially higher. As the island sank, the gulch flooded and the stream deposited sediment into the quiet inlet, filling it to create the flat valley floor.

Hawai'i 440
Lāna'i City–Mānele Bay
7 miles

Hawai'i 440 south of Lāna'i City crosses Pālāwai Basin and follows the Mānele graben to the ocean. Milepost 6 is at the junction where Hawai'i 440 heads west from Lāna'i City. At the harbor in Mānele Bay, Kalaeokahano seacliff exposes many lava flows related to the building of the Lāna'i volcanic shield. On clear days, you can look across the channel to the distant profile of Haleakalā's southwest rift zone on Maui, 30 to 40 miles away.

Hawai'i 440 crosses Pālāwai Basin immediately south of Lāna'i City. Watch for the steep escarpment along its eastern margin. The peak is called Lāna'ihale, and at 3,370 feet, it is the highest point on the island. The low rim on the western horizon is the divide between Pālāwai Basin and Miki Basin.

At milepost 10, 4 miles south of Lāna'i City, the road leaves Pālāwai Basin and reaches a crest with a breathtaking view of the island of Kaho'olawe, 20 miles southeast across Kealaikahiki Channel. On a clear day, you can see the summit of Mauna Kea on the Big Island, 100 miles southeast.

The basaltic shield of Lāna'i may once have towered as high as 9,000 feet. Kaho'olawe, too, was once thousands of feet higher than it is today. Like all these islands, they lost elevation as their weight depressed the ocean floor.

Mānele Graben and Its Tsunami Debris

Look east from the vantage point at the south end of Pālāwai Basin to see a high escarpment rimmed with eroded cinder cones. This is the eastern boundary of the Mānele graben, the strip of sunken land that extends south from Pālāwai to the sea. It was formed when a long slice of the island dropped between parallel faults. The highway switchbacks down the middle of the graben to the coast at Mānele Bay. Watch near milepost 11, on the downgrade south of Pālāwai Basin, for roadcuts in basalt, weathered red. The flows are thin and full of bubbly vesicles, with rubbly bases. Lava flows that channeled their way down the graben built the small peninsula along the western side of Mānele Bay. High cliffs on the east side of Mānele Bay expose rubbly basalt flows of the Lāna'i shield, all dipping seaward.

Hulopo'e Stream eroded the gulch it descends in Mānele graben, west of the highway. Blocks of broken reef rock about 100,000 years old litter the

179

The low peninsula in the distance is a basalt flow that separates Hulopoʻe and Mānele bays.

slopes of Hulopoʻe Gulch to an elevation as high as 1,070 feet above sea level. They are hard to spot from the road, but you can see a good exposure at the Mānele Bay Hotel on the coast. The best place is the small cliff below the tennis courts on the inland side of the hotel. The debris is black basalt rubble with a few small blocks of reef rock. They seem remarkably out of place a hundred feet or so above sea level.

For many years, geologists tried to interpret those blocks as evidence of an old shoreline. But the reef rocks do not assemble into a coherent structure, as an ancient reef deposit should. It is difficult to imagine how a former shoreline could be high on the slopes of an island that appears to have sunk as it moved off the Hawaiian hot spot. Also, surf tends to sort the pieces into areas of different sizes, depending on the size of the fragments and the strength of the waves.

Most geologists believe that a gigantic tsunami washed those blocks of reef limestone up the slopes of Lānaʻi. They suspect that the wave accompanied the ʻAlikā Slide off the west coast of the Big Island, 100 to 120 miles to the southeast. The same wave may also have washed most of the soil off the slopes of neighboring Kahoʻolawe to an elevation of around 800 feet.

Tsunami rubble—blocks of dark basalt and pale reef limestone in a cliff at the Mānele Bay Hotel.

The road down to the golf clubhouse at the Mānele Bay Hotel has good exposures of basalt flows. One massive, black basalt flow, full of holes from gas bubbles, caps a layer of bright red soil about 6 inches thick, on top of black basalt rubble.

At the Shore

Big blocks of basalt in the breakwater in the bay are full of holes, but these are not gas vesicles of the type you normally expect to see in basalt. Look carefully to see that most of them are within the rusty, weathering rind on the basalt, which is less than an inch thick. The fresh rock contains few holes. Salt weathering made these holes. Salt spray soaks into the slightly permeable rock. Then the little crystals of salt that grow in the rock as the water evaporates pry the mineral grains apart. Once a depression forms, it collects salt spray that soon enlarges it into a hole.

It takes a long time for salt weathering to erode those little holes, perhaps hundreds of years. So the intense salt weathering on some surfaces of the basalt blocks in the breakwater probably predates the building of the breakwater. Those blocks came from a shoreline source.

The shore at Hulopoʻe Beach Park, the most popular swimming and picnicking beach on Lānaʻi, is made of calcareous sand washed in from a

Evidence of salt weathering on the breakwater at Mānele Bay.

fringing reef. The many rocky pools and inlets at the east end are part of a Marine Life Conservation District, perfect for snorkeling and diving when the ocean is calm. One large pool was blasted out to make a swimming hole for children.

Hawai'i 430
Lāna'i City–Shipwreck Beach
8 miles

In just 8 miles, this beautiful drive crosses the axial ridge of Lāna'i in the northwest rift zone and descends the island's northern flank to the coast, near the mouth of Maunalei Gulch, the largest canyon on Lāna'i. East Moloka'i and West Maui are across the channels to the northwest.

Pullouts near milepost 1 afford a sweeping view of the Lāna'i tableland. On the western horizon is Pu'u Koa, a broadly rounded landform that may be a lava shield in the northwest rift zone, or perhaps it is a slide remnant. The scarcity of small cones on this part of Lāna'i is a mystery. If this is indeed a rift zone, more vents should be apparent. Lāna'i City, with graceful stands of Norfolk Island pines, appears to the south, as do the Pālāwai and Miki basins.

A deep carpet of red laterite soil blankets the Lāna'i tableland. It originally covered the island all the way to the sea, though it was thin on the slopes and absent in the bottoms of the gulches. Severe erosion that resulted from the introduction of cattle grazing and modern agriculture during the past two centuries has stripped much of the soil from the slopes and has largely destroyed the original grassland and forest. Once the plant cover was

A weathered roadcut near the crest of Hawai'i 430 shows residual stones rounded by deep weathering of a pāhoehoe lava flow. This lava is on its way to becoming soil. —David Saltzer photo

gone, broad patches of infertile ground fell prey to rain-splash erosion and the untiring wind.

A roadcut on the ridge crest 1.5 miles from Lāna'i City exposes pāhoehoe lava flows weathering to laterite soil. These rocks are so deeply weathered that they consist mainly of iron oxide minerals and clay, but you can see the ghostly outlines of their original structure in the soil. Geologists call such material saprolite.

About 2 miles north of Lāna'i City, the view of Moloka'i north across Kalohi Channel is breathtaking. The broad shield of East Moloka'i Volcano is obvious, with the jagged ridgeline corresponding roughly to the position of the caldera. To the west, the low, broad Ho'olehua Saddle links East Moloka'i with its smaller neighbor, West Moloka'i. Cinder cones of late alkalic basalt rise in the saddle area. West Moloka'i was vigorously active 1.5 to 1.9 million years ago; activity on East Moloka'i was slightly more recent.

The small islet off the eastern tip of the Moloka'i is Mokuho'okini, one of the two volcanic vents from the rejuvenated stage of activity.

Look east across 'Au'au Channel to see West Maui rising a mile above sea level. West Maui grew at about the same time as Lāna'i. Many late-stage cinder cones are scattered across the upper slope of the volcano. The cones close to the shore directly opposite Lāna'i are from the rejuvenated stage of activity.

Rounded pāhoehoe lava tongues have been exposed in a roadcut near the crest of a hill. After lengthy weathering, the rocks have turned into clay, called saprolite.

If you can locate the large town of Lahaina along this shore, look upslope to see where a cinder cone blocked a stream as it grew, diverting it around to the eastern side.

Maunalei Gulch and Descent to the Shore

Also east of the road lies Maunalei Gulch, with steep walls extending from the coast to the summit of the island. They are especially high and rugged in the interior range. A communications repeater station perches on the western flank of the canyon.

Upstream, Maunalei Gulch turns from a northeast to a northwest trend. This change in direction aligns the upper reaches of the gulch with the scarps that developed when the Clark Slide dropped the western part of Lāna'i into the ocean. The scarp faults deeply fractured the bedrock, providing easy access for water, which weathered the rock. Then Maunalei Stream preferentially eroded the fractured, weathered rock along the faults.

At lower elevations, a layer of deep orange and red laterite soil appears in cross section at the top of the southern wall of Maunalei Gulch, which is eroded in basalt flows erupted in the shield-building stage of activity. No such deep soil exists closer to the coast, presumably because it was eroded. Perhaps it was lost as a result of overgrazing or—as on Kaho'olawe—to a giant tsunami.

About 4 miles north of Lāna'i City, route 430 passes through an unusual landscape of bedrock pinnacles and colorful soil. This is an area of deep erosion, where you can see weathered rocks stripped of their soil cover. The original laterite soil was at least as thick as the tallest bedrock pinnacle.

Bedrock pinnacles have been exposed by severe soil erosion on the northern slope of Lāna'i.

Similar pinnacles still embedded in soil are exposed in roadcuts along the Kaumalapau highway.

Shipwreck Beach and Reef

A kīawe (mesquite) forest obscures the view of the sea at the end of the paved road. Either of the two dirt tracks will take you to the shore, but the route to the west is more interesting. It leads to Shipwreck Beach, named for several large vessels that over the years have run aground on the island's wide, northern reef. Follow the track west a mile or so to the end. About another half mile west is the rusting hull of an old liberty ship stranded on the shallow reef offshore. It was stranded in 1960. Just short of the end of the dirt track, layered beach rock slopes up onto the edge of the beach.

Fossil burrows are abundant in the layers of coarse, calcareous beach rock that must have accumulated when sea level was higher than it is today. The layers tilt gently seaward, as they do in modern beaches. Fragments of reef debris, including beautiful coral clusters, wash up all along the shore. Beachcombers sometimes find a rare paper nautilus shell or a Japanese glass fishnet float on this beach.

Waves break along the edge of the modern reef several hundred feet offshore. The reef is quite wide here because Lāna'i is sinking; the distance

Beach rock at Shipwreck Beach on the North coast of Lāna'i.

between the reef edge and the shore increases as coral continuously builds up to stay within reach of strong sunlight. If current trends continue, Lāna'i will eventually submerge, becoming one more Pacific atoll delineating the original shape of the vanished island.

Munro Trail, Four-Wheel-Drive Track
Hawai'i 430 – Across Lāna'ihale to Hawai'i 440
About 15 miles

This dirt track is an adventure in motoring. It heads east from near the pass on Hawai'i 430 to wind through dense forest and deep gullies cut in thick, dark clay weathered from basalt. It corkscrews down the northern slopes of the upper part of Maunalei Gulch. Just when you are sure you are lost, it climbs onto the sharp ridge crest of Lāna'ihale, the highest point on Lāna'i. The ridge crest trends southeast, is only a few tens of feet wide in some places, and drops off precipitously on both sides—no place for a timid driver. The views east to Maui and west down to Pālāwai Basin are spectacular. Deep laterite soils are prominent along the ridge. Mudholes that survive long after a rain, even on the ridge crest, show that the red clay of the laterite is nearly impermeable. From Lāna'ihale, the track descends southward, then westward through progressively drier forest until it reaches the southern edge of Pālāwai Basin. The road improves in the low open country along the southern edge of the basin and finally reaches Hawai'i 440 north of Mānele Bay.

Polihua Road to Garden of the Gods, Four-Wheel-Drive Track
Lāna'i City – Polihua Beach
10.5 miles

Polihua Road is a dirt road that leads west from Hawai'i 430 about 0.3 mile north of Kō'ele Lodge. Set your odometer to zero at the highway intersection. In a little over half a mile, the road abruptly turns northwest, then heads essentially straight to Garden of the Gods, 6 miles away. For several miles, the road crosses nearly flat range land. The low ridge to the north may be the old northwest rift zone of Lāna'i. A line of Norfolk Island pines was planted along it years ago in an attempt to make rain by slowing the clouds in their passage across the island.

The road climbs gently into a dense forest of ironwood trees, which have feathery needles more than 6 inches long. Watch about 5.4 miles from Hawai'i 430 for a picturesque patch of residual stones standing like mushroom caps on stems of red laterite soil. The pedestals form because the soil beneath is protected from rain-splash erosion.

After about 6 miles, the road comes to a gate at the entrance to the Garden of the Gods, a wonderland of residual stones that stand like frozen statues in a sea of red laterite soil. The rounded residual stones are remnants of weathered basalt that erupted in the northwest rift zone. You can see how they form by looking carefully at the base of the emerging boulders where they rest on the laterite soil. Concentric weathering rinds of red clay a quarter to half an inch thick mark the transition from the less weathered residual stone to the deeply weathered laterite. In some places you can see concave, scalloped cusps in the faces of the boulders where weathering has penetrated deeper. Water permeates the outer half inch or so of the less-weathered rock. It dissolves sodium and calcium from the minerals in the rock, converting them to clays rich in aluminum and silica. Even the silica finally dissolves, leaving the red laterite. Iron oxides remain to the end, providing a spectrum of red, yellow, and brown.

Nature did not create the stacks of the rocks on top of the large residual stones. Visitors and islanders do that, setting up small traditional shrines.

A short distance northwest, where the track to the right drops into a more deeply eroded valley, you can see the depth of lateritic weathering in the thickness of the red soil. Laterite is thinner downslope, where some gray unweathered basalt is exposed. Polihua Beach is a strand of beige sand eroded from the coral reefs offshore.

Mushroom-cap residual stones are weathering out of vesicular basalt about one-half mile east of Garden of the Gods.

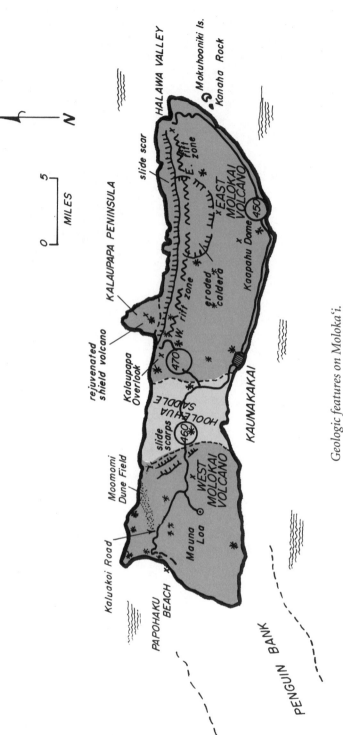

Geologic features on Moloka'i.

5
Moloka'i, The Friendly Isle

Moloka'i is the oldest member of the Maui Nui group of Hawaiian Islands: Moloka'i, Maui, Lāna'i, and Kaho'olawe. They formed a single land mass during the latest ice age, when sea level was low enough to drain the narrow straits now separating them. At least six, and possibly seven, coalescing shield volcanoes make up the group.

Moloka'i has an area of 260 square miles; it is 38 miles long and only 10 miles wide. A bird's-eye view suggests a deformed peanut, with shield volcanoes at both ends. East Moloka'i is the younger. West Moloka'i is a small volcano, long quiet.

West Moloka'i Volcano

West Moloka'i built above the waves around 1.9 million years ago. Like other Hawaiian volcanoes, it erupted fluid flows of basalt during the initial period of rapid growth, constructing a broad, low-profile shield. Rift zones developed along trends northwest and southwest from the summit. If the volcano ever had an east rift zone, any evidence of it either was buried beneath the younger lavas of East Moloka'i or went to the deep ocean floor in a slide, perhaps together with the summit and caldera of West Moloka'i.

As West Moloka'i drifted west off the Hawaiian hot spot, the composition of lavas in it changed from the ordinary tholeiitic basalts of the main stage of shield volcanism to the more exotic alkalic basalts of late-stage activity. The late eruptions added sixteen cinder and lava cones to the landscape, and several alkalic lava flows. Most of these eruptions were in the northwest rift zone. West Moloka'i shows no rejuvenated-stage volcanism.

A high submarine platform stretches about 40 miles southwest from West Moloka'i and culminates in Penguin (Penquin) Bank, which covers about 380 square miles and lies at a water depth of only 170 feet. The flat top on the platform may be from a combination of wave erosion and reef growth when sea level was lower, presumably during one or more recent ice ages. Some geologists speculate that Penguin Bank is an offshore extension of the West Moloka'i southwest rift zone. Others suggest it may be a separate, submerged shield volcano topographically linked to West Moloka'i by a submarine extension of the southwest rift zone.

During or shortly after the late stage of alkalic volcanism, the summit and northeastern flank of West Moloka'i collapsed into the ocean. Their departure left a set of large slide scarps across the sundered top of the mountain. Flows from neighboring East Moloka'i built up against these scarps, showing that it is a much younger volcano.

West Moloka'i is dry. It is too low to snatch moisture from the clouds, and East Moloka'i shelters it from the big storms that come in on the northeast trade winds. Stream erosion has minimally modified the volcano since it went out of business.

East Moloka'i Volcano

Age-dating tests show that East Moloka'i Volcano was building above sea level as early as 1.75 million years ago. As East Moloka'i grew, lava flows from it piled up against the flank of West Moloka'i, building the broad, nearly level Ho'olehua Saddle between them. Several feet of red laterite soil had developed on the lavas of West Moloka'i before the eruptions from East Moloka'i covered them. So it seems that the West Moloka'i Volcano was thoroughly dead for some time before East Moloka'i reached its full size.

East Moloka'i had two rift zones: one extended west of the summit, and another reached eastward. A caldera 5 to 6 miles across opened in the summit, which may once have stood 11,000 feet above sea level.

In many places, a prominent laterite paleosol separates the lower lavas of the shield-building stage from the alkalic lavas of late-stage activity. In other places, notably in the western rift zone, the two kinds of flows are interbedded. Apparently, the change from one type of volcanism to the other followed a long pause in activity in some areas, but not in others.

Radiometric age dates on the lava flows show that the change from shield to late-stage volcanism was between 1.5 and 1.4 million years ago. The late-stage eruptions built at least a dozen cinder cones and domes of trachyte and covered the shield with lava flows of alkalic basalt. Many streams have since cut through this veneer of late-stage lava, exposing the shield basalts beneath.

Wailau Slide

As late-stage alkalic volcanism waned, or shortly after it ceased, the northern flank of East Moloka'i broke approximately along its two rift zones. Half of the volcano disappeared into the ocean in the Wailau Slide. It must have been moving fast, because it shattered into fragments that spread nearly 100 miles north across the Hawaiian Deep, in a strip 25 miles wide.

The Wailau Slide is the third largest slide identified in Hawaiian waters, after the North Kaua'i and Nu'uanu slides. James Moore of the U.S. Geological Survey estimated that the volume of this slide corresponds to the amount of lava the Hawaiian hot spot would erupt in 10,000 years, at its current rate.

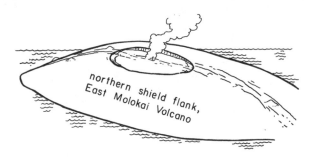

northern shield flank,
East Molokai Volcano

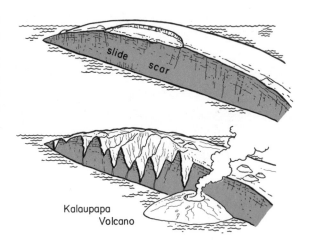

The Wailau Slide removed the northern flank of East Moloka'i. Later, a rejuvenated eruption built a small shield volcano against the slide scar at Kalaupapa.

slide scar

Kalaupapa
Volcano

The Wailau slide split the caldera of East Moloka'i in half. The scar it left behind is the towering sea cliff along the island's north shore, the highest and quite possibly the most spectacular shoreline cliff in the world. It rises almost vertically 3,700 feet out of the sea. Islets and sea stacks along the base show that waves have eroded the cliff at least a short distance inland.

Large stream valleys have cut across the cliff face to a depth of nearly 6,000 feet below sea level. They were eroded when the island stood much higher, and became submerged as it sank. Obviously, the great cliff was then 6,000 feet higher, too.

Rejuvenated Volcanic Activity

Around 300,000 years ago, a series of rejuvenated eruptions built a small volcanic shield, the Kalaupapa Peninsula, against the western base of the East Moloka'i sea cliff. Most of the alkalic basalt flows in the peninsula are ropy pāhoehoe with glassy surfaces. They contain olivine.

Kauhakō Crater is at the summit of the Kalaupapa shield, nearly 400 feet above sea level. It is a combination explosion and collapse pit about a thousand feet across. Groundwater and surface runoff maintain a pond at the bottom. A huge lava channel and tube system extend north from Kauhakō Crater along the length of the peninsula. Pāhoehoe lava flows spilling from this channel built the surrounding land.

A tsunami deposit at the northeastern edge of Kaunakakai contains white pieces of reef material and chunks of black basalt. View is about 14 inches across.

Another eruption during the rejuvenated phase of volcanic activity built a small ash and lava cone near the eastern cape of Molokaʻi. It has since eroded into tiny Kanahā Rock and larger Mokuhoʻoniki Island.

Kaunakakai and Deposits from a Giant Tsunami

Kaunakakai, the main seaport and population center of Molokaʻi, lies near the center of the south coast of the island. The center of town is close to sea level; newer residential areas are creeping up the gentle flanks of East Molokaʻi Volcano. A low, rusty cinder cone, Puʻu Maninikolo, on the northeastern edge of town, has been partly excavated for road gravel.

Much of Kaunakakai is built on deposits from an ancient giant tsunami, perhaps the one generated by the collapse of the ʻAlikā Slide off the west coast of the Big Island 100,000 years ago. The deposits are easily seen in residential areas south and west of the cinder cone in the northeastern part of town.

Watch the excavated edges of residential streets for exposures of angular fragments of white reef rock and basalt ranging from an inch to 2 feet across. The soil cover is thin, so the same deposits appear in most excavations. Look carefully for coral and small shells. The best exposed and most accessible outcrops are within about 100 feet above sea level, although some are at higher elevations. The island has been sinking as it moves off the Hawaiian hot spot, so those deposits were considerably higher above sea level when the great waves laid them down.

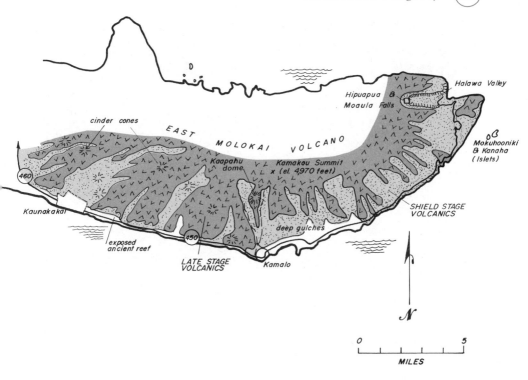

Hawaiʻi 450, Kamehameha V Highway
Kaunakakai–Hālawa Valley
21 miles

Hawaiʻi 450 follows the dry, southern coast of Molokaʻi from Kaunakakai, the main town, to the eastern cape of the island. For the first half of the distance, it stays close to sea level, passing pandanus swamps and shallow bays sheltered behind a fringing reef. Near the eastern cape, Hawaiʻi 450 winds in much diminished form onto the wet side of the island. It ends in Hālawa Valley, where you can see dramatic interior topography typical of the East Molokaʻi windward slope.

Many of the coastal rocks on Molokaʻi record fluctuations in sea level, as do those on the other older Hawaiian Islands. Watch for the roadcut along the north side of Hawaiʻi 450 about a mile east of the main intersection in Kaunakakai. It exposes a reef deposit 3 or 4 feet thick with a flat base on a rubble of basalt. The reef is dirty on exposed surfaces, white where freshly broken.

Molokaʻi is approaching floating equilibrium with the mantle and is no longer rapidly sinking. The vertical stability of the island makes it reasonable to argue that the ancient reef deposits date from when sea level was higher

195

This fossil reef deposit is a mile east of Kaunakakai, on the inland side of Hawai'i 450.

Alkalic basalt shows flow banding and contains xenoliths of gabbro near milepost 8 on Hawai'i 450. —Ronald K. Gebhardt photo

The knob on the horizon is Kaʻāpahu, a trachyte dome.

than today. Presumably, that was during an interglacial period, when the climate was warmer and more land ice was melted.

The bedrock in the area within 4 miles east of Kaunakakai is alkalic basalt that erupted late in the volcanic life of East Molokaʻi. If you look high on the slope near milepost 6, you can see a subdued group of young cinder cones profiled against the sky. They erupted during late-stage volcanic activity.

At milepost 8, a roadcut on the north side of the highway exposes glassy alkalic basalt with conspicuous flow banding. The flow erupted from a vent a short distance up the slope. It contains a few small, weathered xenoliths, probably fragments scooped from crystallized portions of the magma chamber below. A younger ʻaʻā flow covers the eastern margin.

If you look upslope near milepost 9, several miles inland you can see Kaʻāpahu Hill. From this vantage point, it is the highest knob at the left edge of the skyline, a large trachyte dome that oozed downslope more than a mile as an enormously thick lava flow. Trachyte lava is far more viscous than basalt lava, so it tends to make domes and thick flows. Streams have eroded the deep gullies flanking the flow. The ridgeline above Kaʻāpahu culminates in Kamakou Peak, the highest point on Molokaʻi, at 4,961 feet.

Red Laterite and Black Basalt

Between 18 and 19 miles east of Kaunakakai, watch in the upper half of roadcuts for red laterite soil full of rounded residual stones of dark basalt. A few hundred yards east of mile 19 a lava flow covering perhaps the same bright red laterite shows spheroidal or concentric weathering. This same ancient soil appears in many other roadcuts.

Moku Ho'oniki Island and Kanahā Rock.

The lavas below the ancient soil level are tholeiite basalt that erupted during the main shield-building stage of volcanic activity. Lavas above it are alkalic basalt that erupted late in the life of the volcano. The paleosol must represent many thousands of years of weathering, indicating that quite a significant period of time elapsed between the end of the shield eruptions and the first appearance of alkalic lavas. The fringing coral reef lies a couple of hundred yards offshore, at the line of breakers.

Mokuho'oniki Island and Kanahā Rock come into view at milepost 20. They are remnants from one of two known eruptions on Moloka'i in the rejuvenated phase of volcanic activity. A pullout near this milepost permits close examination of the ancient red laterite soil and the flows above and below it. The basalt flow above is massive and black; the ancient soil covers broken rock, either slide rubble or a mudflow, and a lava flow that contains about 40 percent plagioclase in pale crystals as much as an inch long. Similar basalt in other Hawaiian shield volcanoes was erupted during the waning phases of shield-building activity.

The red buried soil continues for another mile or two, typically capped by a massive flow of black basalt. Picturesque little pocket beaches between rocky points of black basalt are beige coral sand eroded from the offshore reef.

Across Halawa Bay to the north, the main shield-building basalt flows of Moloka'i slope gently seaward.

Between miles 22 and 26, the road climbs inland to grassy range-land underlain by rusty lateritic soil. It winds around the upper edge of the slope through a forest of ironwood trees. On one inside bend, a giant ban-yan tree with huge spreading roots drapes an umbrella of dark green leaves over the road.

Near mile 26, Hālawa Valley lookout provides views north across Hālawa Bay and east to the tall sea cliff, Kapaliloa, reaching out to Cape Hālawa at the easternmost end of the island.

As the highway rounds the eastern cape, it passes into the windward rain belt. Roadcuts in the forest expose intensely weathered soil with a much lusher plant cover than farther west. The road finally twists down to the floor of Hālawa Valley, where it leads out to the mouth of the stream. If the weather is clear, you can see two spectacular waterfalls about 2 miles inland. A foot trail leads to the higher one.

Farmers used to maintain many kalo (taro) patches on the flat floor of Hālawa Valley, but tsunami wiped out their fields in 1946 and 1957. Swim-ming from the beaches at the head of the bay is usually fine, though the water is sometimes murky. This is a popular surfing area.

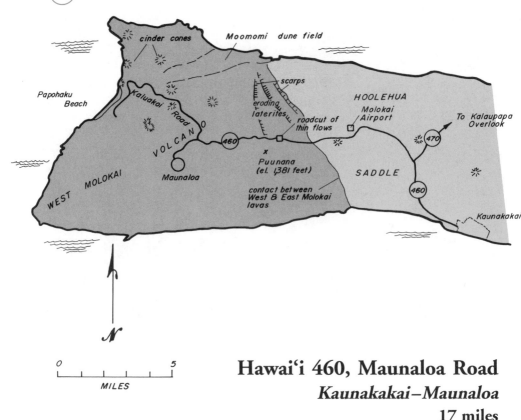

cinder cones

Moomomi dune field

scarps

HOOLEHUA

Papohaku Beach

Kaluakoi Road

eroding laterites

roadcut of thin flows

Molokai Airport

To Kalaupapa Overlook

470

460

V O L C A N O

x
Puunana
(el. 1,381 feet)

SADDLE

460

Maunaloa

WEST MOLOKAI

contact between West & East Molokai lavas

Kaunakakai

N

0 — 5
MILES

Hawai'i 460, Maunaloa Road
Kaunakakai–Maunaloa
17 miles

Hawai'i 460 follows an inland route across Ho'olehua Saddle and the low shield volcano of West Moloka'i. This is open country, more so than two centuries ago when scrubby woodland covered it. The land was cleared for cultivation and grazing.

Wherever cattle graze an area faster than grass grows, they soon strip the ground bare. Rainwater falling on bare soil runs off the surface instead of soaking in. In many places broad swaths of red laterite soil on West Moloka'i are now exposed in intricately eroded terrain of badlands, a sobering monument to land abuse.

West of Kaunakakai, the highway climbs the low flank of West Moloka'i Volcano. Black basalt is exposed in a roadcut 1.4 miles west of town.

Just south of the junction with Hawai'i 470, you can look directly upslope about 3 miles to see a prominent cinder cone of late alkalic basalt, Pu'u Luahine. Such cinder cones typically form when basalt magma rising to the surface on the lower flank of a volcano encounters water in the ground. The water bubbles into the magma, flashes instantly into steam, and, together with volcanic gas, blasts fragments of solidifying bubbly magma into the air. These fragments pile up around the volcanic vent, building a cinder cone.

Residual stones are being eroded out of a bank of lateritic soil on West Moloka'i.
—Ronald K. Gebhardt photo

Ho'olehua Saddle

Hawai'i 460 crosses Ho'olehua Saddle after the intersection with route 470, the road to Kalaupapa Overlook. Like other broad passes between Hawaiian volcanoes, Ho'olehua Saddle formed where the lava flows from a younger volcano lapped against the flank of an older one. Shield basalts and later flows of alkalic basalt that erupted in the west-to-northwest rift zone of East Moloka'i make up most of this saddle. They lap onto the older West Moloka'i lava flows at its western edge.

The rocks of Ho'olehua Saddle are deeply weathered into laterite stained red with iron oxide. Watch for scattered residual stones, especially near the edges of plowed fields. They resemble stream-rounded rocks even though water has not transported them. The residual stones are on the surface because erosion has removed the soil in which they formed and were embedded.

About half a mile west of the junction with Hawai'i 470, and less than a mile directly upslope from near milepost 5, is another small cinder cone, Kualapu'u.

At mile 6.3, look south for a spectacular roadcut that has exposed thin pāhoehoe lava flows weathering into spherical residual stones. You can see similar exposures a short distance up the hill.

An Ancient Broken Summit on West Moloka'i

Hawai'i 460 nears the flank of West Moloka'i near milepost 9, 4.7 miles west of the intersection with Hawai'i 470. You will see many rounded residual stones along the margin of the field north of the highway. The steep slopes

201

Eroded laterite wasteland covers the northeastern flank of West Moloka'i Volcano.

are eroded scarps at the headwall of the massive landslide that dropped much of the northeast flank of West Moloka'i into the ocean. Most of the mass moved underwater, but the slide scar extended well above sea level.

Watch for deep laterite soil exposed in roadcuts where the highway begins climbing the flank of West Moloka'i Volcano. A small cinder cone of late alkalic basalt perches on the crest of the scarp to the south. This is where the projections from the northwest and southwest rift zones of West Moloka'i intersect, indicating that the vanished ancient summit of the volcano must have stood nearby.

A large roadcut towers above both sides of the highway near milepost 12 west of Kaunakakai. The rocks show remarkably regular, horizontal color bands, each one a foot or two thick. Most consist of pale gray ledges, the solid interiors of lava flows, separated by zones of darker and more easily eroded flow rubble. This appears to be a substantial stack of very thin lava flows.

The flows erupted late in the main shield-building stage of the volcano's growth. They are thin because they were especially hot and very fluid. The small dikes cutting vertically across the roadcut indicate that this site lies close to a vent area in the northwest rift zone.

Several prominent, yellowish brown to red zones a few inches thick between flows are baked fine ash and flow rubble. Each successive lava flow heated and altered the underlying wet ash, converting the iron in the upper

Near milepost 12 a stack of thin lava flows is separated by layers of buried soil beds and baked breccias.

layer to red iron oxide, while leaving the less baked ash below yellow. One especially conspicuous red zone 10 to 15 feet high in the central and eastern part of the cut is an ancient soil, a paleosol, as much as a foot thick. Farther west another mile or so are big roadcuts in rubbly black basalt that has deeply weathered at the upper levels to dark orange laterite. Beyond this point you will see more thin basalt flows weathered red.

Moʻomomi Dunes and Road's End

Farther west, Hawaiʻi 460 passes onto a broad tableland that slopes gently seaward. The Moʻomomi Dunes are a large field of pink calcareous sand that stretches inland from the northern coast. All you can see from the road is their margin, now stabilized under plant cover.

The main part of the Moʻomomi Dunes probably was formed during the latest ice age, when sea level was low and the reefs now submerged offshore were dry and feeding sand into the wind. Since then, slightly acidic rain has cemented some of the sand into hard limestone. Paleontologists have found the bones of several extinct creatures, including a flightless goose, in the old dunes.

The road ends at the quaint little community of Mauna Loa, an old pineapple plantation town from the 1920s. These days it is growing some trendy new shops.

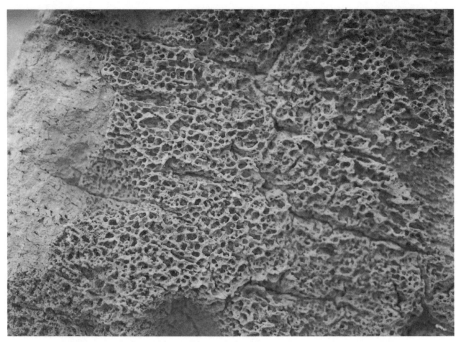

Salt weathering eroded these holes in the surface of a basalt boulder.

Kaluako'i Road
Hawai'i 460 – Pāpōhaku Beach
6 miles

From Hawai'i 460, Kaluako'i Road leads to the shore just south of 'Īlio Point, the northwestern cape of Moloka'i. Trees along the road lean sharply to the southwest, parallel to the belt of sand dunes in the Mo'omomi field to the north. Both indicate the prevailing direction of strong storm winds.

Late-stage cinder cones and lava shields made of alkalic basalt that erupted along the northwest rift zone are north of the road and around Kaluako'i Resort. One of these, a prominent dark red cinder cone, Okoli, stands at the north side of the road 2.6 miles west from Hawai'i 460. The east and north sides of the cone have been excavated for road gravel.

Kaluako'i Road leads to Pāpōhaku Beach, on the western end of Moloka'i. Two miles long, this is one of the longest calcareous sand beaches in Hawai'i. Without a fringing offshore reef to provide a local supply of sand, it seems likely that the steady northeast trade winds blow sand in from the Mo'omomi Dunes.

The lack of an offshore reef also leaves the coast with nothing to absorb the force of the ocean waves. The offshore slope is very steep; waves do not meet shallow water until they come crashing onto the beach. This is a treacherous place to swim.

Farther south, public access spurs off the main road lead to several little pocket beaches between rocky basalt points. Watch for basalt boulders pockmarked as if with gas holes, especially at the southernmost beach, Kapukahehu. No such holes pock the recently broken surfaces, so they cannot be an original feature of the lava flow, as the gas holes are. These holes are the result of salt weathering.

Salt spray soaks into the surface of the rock, then little salt crystals form as it dries. The crystals growing in the rock pry little chips from the surface. More salt spray concentrates in the depressions, localizing the weathering and deepening the holes.

On clear days, you can see windward O'ahu, 25 miles across the Kaiwi Channel. The profiles of cones that erupted in the Koko Rift between 32,000 and 6,000 years ago rise along the left side of the island's profile.

O'ahu stands on the same general submarine platform as Moloka'i, Maui, Lāna'i, and Kaho'olawe, but the Kaiwi Channel is deeper than the channels between the other islands. O'ahu may once have been linked to Moloka'i, but isostatic subsidence has long since separated them.

Hawai'i 470
The Road to Kalaupapa Overlook
6 miles

Hawai'i 470 heads north about 6 miles from Hawai'i 460, the main east-to-west road across Ho'olehua Saddle. A couple of miles east of the highway, north of the intersection, is a prominent rounded knob. This is Pu'u Luahine, an old, rusty cinder cone. After passing the community of Kualapu'u, the road quickly climbs the western flank of East Moloka'i Volcano into lush, green uplands, skirting several late-stage cinder cones near Kala'e.

On a clear morning, you can look west about 10 miles in the stretch between mileposts 4 and 5 to see the beige patches of the Mo'omomi Dunes along the northern coast, and the slide scarps that cut across the northeastern flank of West Moloka'i.

A clear morning also offers the best chance of good views from the Kalaupapa Overlook, at the end of the road. You can look down 1,600 feet onto the Kalaupapa Peninsula. A volcanic afterthought, it is a small shield volcano dabbed onto the north shore of Moloka'i at the base of gigantic sea cliffs. No road descends the wall. Lack of easy access helps explain why Kalawao, on the eastern side of the peninsula, was selected as a leprosy reserve by the

The view eastward from Kalaupapa Overlook. The 3,000-foot cliffs along the north Moloka'i coast vividly emphasize the giant landslide scar left when half the island slumped into the ocean.

A 1,700-foot cliff towers above 'Awahua Bay, across from Kalaupapa Peninsula. —David Saltzer photo

Kingdom of Hawaii, and later by the U.S. government. Now that modern drugs control leprosy, the reserve is a National Historical Park open to visitors.

Hawai'i 470 ends at a pair of small parking lots in Pālā'au State Park. A short trail from the east lot leads to a lookout with a spectacular view. The view to the east is across 'Awahua Bay. Kalaupapa Peninsula is the low tongue protruding from the base of the towering cliffs. Kauhakō Crater is clearly visible near the middle of the peninsula; if the light is good, you can see the long lava channel leading from it across the shield's northern slope. Beyond lie Mōkapu and Ōala islands, at the base of the cliff.

With permission obtained through the Moloka'i Visitors Association, you can visit the peninsula and the former leprosarium by boat, by airplane, or by following a difficult 3-mile trail down the 1,700-foot cliff. Mule trains are sometimes available to take groups of people down the trail. Watch from the trail and from the roads below for several dikes, vertical ribbons of black basalt, exposed in the cliff face. They are part of the western rift zone of East Moloka'i, exposed when the gigantic Wailau Slide parted company with the rest of the volcano and plunged into the ocean.

From the west parking lot, at the end of the road, a short trail leads through a forest of ironwood trees to a small rock formation called Ka Ule o Nānāhoa (Phallic Stone). This was a worship site for early Hawaiians. Although modified by hand carving, it began as a residual stone like the other residual stones nearby that show concentric weathering rinds.

The rock is a gray trachyte that evidently differentiated from basalt magma during late-stage volcanic activity. It contains an abnormally high concentration of sodium. The extremely viscous trachyte lava erupted as a dome.

Mōkapu Island from near the rim of Kauhakō Crater on the Kalaupapa Peninsula. —David Saltzer photo

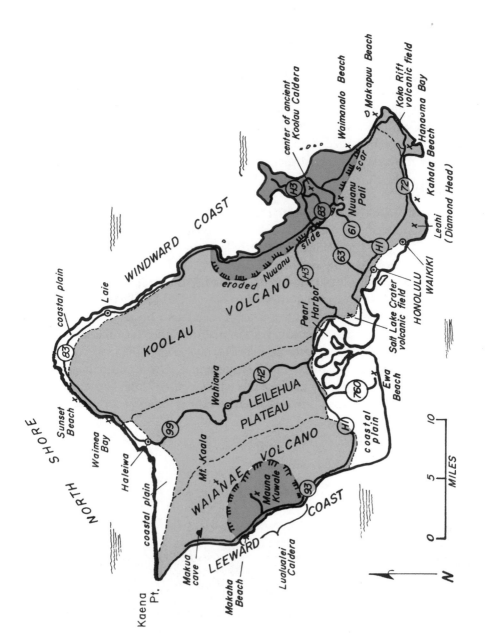

Major geologic features on O'ahu.

6
O'ahu, The Gathering Place

O'ahu, the political and economic heart of Hawai'i, packs a large population into an area only 44 miles long and about 30 miles wide, 604 square miles. Two rugged mountain ranges, both deeply eroded shield volcanoes, form the eastern and western regions of the island. Between them lies a broad alluvial plain, the Leilehua Plateau, which stretches south to meet the sea at Pearl Harbor and north to the famous big-wave surfing beaches along the North Shore. Early Hawaiians used Pearl Harbor and the neighboring coastal plains as a gathering place on the occasion of important religious festivals and other events.

Wai'anae Volcano

O'ahu began as a small, vigorously active volcanic islet that first rose above sea level nearly 4 million years ago. During the next million years, the young island grew rapidly as lava flows built the huge shield of Wai'anae Volcano. Its eroded remnants are the Wai'anae Range, which includes Mount Ka'ala, the highest peak on O'ahu (4,025 feet). The conspicuous flat summit of Ka'ala is the high point on the skyline west of Honolulu. This mountaintop flatland, a moist and nearly inaccessible bog, is one of the few remnants from the original surface of Wai'anae, the oldest terrain on the island.

The widely recognized profile of Diamond Head looms over Waikīkī Beach.

Like other young Hawaiian volcanoes, the Waiʻanae shield had a caldera at the summit, about 4.5 miles across. Two narrow, active rift zones extended northwest and southeast from the caldera. Swarms of dikes across them constitute the resistant backbone of the Waiʻanae Range. A dispersed zone of radiating dikes also shows evidence of flank eruptions northeast of the caldera.

The overwhelming bulk of Waiʻanae Volcano is thin flows of pāhoehoe lava, tholeiite basalt. These oldest rocks built the main shield between 4 and 3 million years ago, when the volcano was over the Hawaiian hot spot. Since then it has moved about 180 miles northwest. The flows dip outward from the caldera at angles ranging from less than 5 degrees along the rift zones to almost 15 degrees in the other flanks. Geologists call them the Lualualei member of the Waiʻanae formation.

As Waiʻanae Volcano completed its massive shield, eruptions became less frequent in the rift zones and were mainly restricted to the summit. At first, thick tholeiitic basalt flows ponded in the caldera. Later, the flows became more alkalic and viscous. Geologists call these late lavas the Kamaileʻunu member of the Waiʻanae formation.

By the time the caldera was filled, Waiʻanae was erupting mainly alkalic basalt flows, although some vents produced cinder, ash, and other pyroclastic material. Throughout the period of caldera filling, the outer flanks of Waiʻanae eroded, so the alkalic lava flows buried a landscape laced with young stream channels and small valleys. By about 2.9 million years ago, alkalic flows and ash covered most of the Waiʻanae shield. The eruption of these late-stage alkalic rocks marked the end of the volcano's main growth period. They are the Pālehua member of the Waiʻanae formation.

While Waiʻanae was still growing, its western submarine slope began slowly slumping into the Hawaiian Deep. Eventually, most of the shield's western flank slid into the ocean, spreading chunks of volcano across the sea floor as far as 50 miles from the island. The north flank of Waiʻanae was also unstable. The Kaʻena Slide moved nearly 70 miles across the deep ocean floor and formed an underwater escarpment parallel to the North Shore between Kaʻena Point and Waialua. It is difficult to imagine how slide debris could move so far across the ocean floor unless it was moving extremely fast. This slide may have been a catastrophic event, the kind that generates enormous tsunami.

Percolating gases, steam, and hot water altered the lava flows inside Waiʻanae caldera. The weakened rock weathered and eroded rapidly, opening a broad flatland, Lualualei Valley, which extends inland from the town of Waiʻanae. Erosion also stripped most of the alkalic lavas off the northern flanks of the volcano and exposed dike swarms deep within the rift zones. Thick deposits of stream gravel, conglomerate, and other alluvium accumulated in the growing valleys and lowlands.

An episode of rejuvenated volcanism at the southern end of the Waiʻanae Range built six small cinder and spatter cones, which poured flows of alkalic basalt onto the southeastern flank of the old shield, interrupting the erosional dissection of the volcano. Radiometric age dates show that these volcanic rocks, the Kolekole basalts, erupted between 2.9 and 2.75 million years ago.

Strictly speaking, the Kolekole volcanic activity was not rejuvenated volcanism, because little time intervened between the eruptions of Pālehua and Kolekole basalt. Both probably erupted from the same magma source, but a brief interval of erosion separated the two phases of activity. Some geologists believe the huge Waiʻanae slump interrupted the volcano's late-stage eruptions. If so, then Waiʻanae Volcano never had a spell of rejuvenated activity.

Koʻolau Volcano and the Great Nuʻuanu Slide

Koʻolau Volcano is younger and larger than Waiʻanae, and accounts for about two-thirds of Oʻahu. It grew above sea level by 2.7 million years ago, about when Waiʻanae Volcano was becoming extinct. The first stage of growth for Koʻolau was the usual rapid eruption of pāhoehoe basalt flows. A caldera nearly 8 miles long and at least 4 miles wide developed in the present coastal lowland between Kailua and Kāneʻohe. Two vigorously active rift zones trended northwest and southeast of the caldera, roughly parallel to the rift zones on Waiʻanae.

Koʻolau and Waiʻanae may have grown as separate islands. The broad saddle of the Leilehua Plateau that connects them consists mainly of alluvial deposits shed from Waiʻanae and lava flows that came from Koʻolau. Koʻolau shed more alluvium onto the saddle as volcanic activity waned.

Nuʻuanu Pali on the windward coast of Oʻahu, near Kāneʻohe.

The Leilehua Plateau and Wai'anae braced the western flank of Ko'olau Volcano, while the unsupported eastern flank sloped steeply into the deep ocean. During the stage of most rapid volcanic growth, this steep flank on Ko'olau swelled as magma rose into the volcano, then deflated as it erupted. Those movements probably helped destabilize the eastern flank enough to cause the Nu'uanu Slide.

The Nu'uanu Slide is one of the largest known slides on earth. It sped across the sea floor, carrying chunks of Ko'olau as far as 120 miles to the northeast. The slide laid down a bed of rubble 20 miles wide. One block is 18 miles long and a mile thick, as large as many of the ancient seamounts on the surrounding ocean floor. The slide must have had tremendous momentum to run all the way across the Hawaiian Deep and 90 miles up the gentle slope of the Hawaiian Arch. It may have caused a horrendous tsunami all around the Pacific Rim.

You can see one legacy of the Nu'uanu Slide in the way the eastern shore of O'ahu almost follows the Ko'olau rift zones and crosses the eastern fringe of its caldera. Had the Nu'uanu Slide not detached a large piece of Ko'olau, the eastern shore would lie several miles seaward.

About 1.8 million years ago, probably not long after the Nu'uanu Slide, Ko'olau entered its last gasp of activity. It erupted almost no alkalic basalts after shield volcanism ended. During the next 800,000 years, the volcano eroded and sank thousands of feet as it drifted away from the Hawaiian hot spot. Erosion carved broad valleys out of what remained of the caldera floor after the Nu'uanu Slide, exposing countless dikes in the rift zones.

A vigorous episode of rejuvenated volcanism then interrupted the slow decay of Ko'olau and produced the Honolulu basalts. The first eruptions began a million to 850,000 years ago on Mōkapu Peninsula, at the northern margin of Ko'olau caldera. Since then, the southern part of the range has had about forty more eruptions. The largest group of Honolulu volcanoes trends south from the Mōkapu Peninsula across the range crest to the coastal plain at Honolulu. It includes Punchbowl, the site of a military cemetery, Pu'u Kāhea (Sugarloaf), Pu'u 'Ōhi'a (Tantalus Peak), Pu'u 'Ualaka'a (Roundtop), and Lē'ahi (Diamond Head). A lesser cluster of vents that includes Salt Lake Crater, Makalapa, and Āliamanu cones rises from the coastal plain between Honolulu and Pearl Harbor. The youngest cluster of pyroclastic cones erupted between 32,000 and 6,000 years ago in a chain that extends northwest from Makapu'u Point and Koko Head, at the southeastern end of the island.

The Honolulu basalts include some exotic varieties of basalt unusually rich in sodium. They include nephelinite, a lava that contains the rare mineral nepheline, and basanite, which is not as strongly alkalic and contains ordinary plagioclase, along with nepheline. Some lava flows contain xenoliths of lherzolite and pyroxenite, mantle rocks carried to the surface as fragments. They provide geologists with a view of the earth's interior beneath Oʻahu.

In *Volcanoes in the Sea*, Gordon Macdonald, Agatin Abbott, and Frank Peterson discuss the rejuvenated volcanism of Koʻolau: "The intervals of time between successive eruptions seem to have been as long as, or even longer than, the length of time from the last eruption to the present" (93–94). If they are correct, perhaps we can expect future eruptions.

Oʻahu Shores

The geologic history of Oʻahu is more than the rise and decline of its two volcanoes. Beautiful coral reefs and coral sand lagoons in the warm tropical water have also played a role.

Oʻahu is no longer sinking into the ocean, perhaps because the mantle below has finally adjusted to the load. Or perhaps, as some geologists believe, the ocean floor may be flexing to compensate for the weight of the younger islands growing to the southeast, thus offsetting the island's sinking. Whatever the reason, Oʻahu has the most stable coasts in the Islands.

Even though Oʻahu is not sinking, the shorelines show such evidence of recent fluctuations in sea level as drowned stream valleys, dead reefs and lagoons above sea level, ancient beach rubble, and beach rock. Much of that evidence records the coming and going of ice ages. Sea level drops as much as several hundred feet when glaciers grow on the continents, then rises as all that water drains back into the ocean when they melt.

Waves break in the shallow water of Pūpūkea Beach, North Shore of Oʻahu.

H-1 Freeway
Honolulu – Farrington Highway
20 miles

West of downtown Honolulu, the H-1 Freeway and Hawai'i 78 follow the margin of the Salt Lake Crater volcanic field. The freeway passes south of Salt Lake and Āliamanu craters, but traverses the marshy center of Makalapa Crater. Hawai'i 78 skirts the northwestern edges of Salt Lake and Āliamanu Craters. The two highways merge at 'Aiea.

The Salt Lake Vents

The Salt Lake group of volcanoes features low ash cones with broad rims and wide craters that have filled with lake and pond sediments. Their shape is typical of vents that erupt explosively after rising magma encounters groundwater at shallow depth and flashes it into steam. Blocks of basalt embedded in the ash deposits show that the steam explosions were powerful enough to tear chunks out of the older rock and blow them out of the vent along with the ash. The soft layers of ash sagged beneath their weight. In places, surges of steam blowing out of the vents worked the ash deposits into crossbedded dunes.

The Salt Lake volcanic field includes at least six ash cones. The largest is Salt Lake Crater, 1.4 miles across. Makalapa Crater stands alone at the western end of the cluster. The others overlap, and may have formed during the same large eruption.

No one now knows when the Salt Lake vents erupted. Rocks suitable for dating are difficult to find because the volcanic pile contains no lava flows, and the ash is altered and weathered. Geologist Harold Stearns attempted to correlate the Salt Lake ash with fossils from drill holes in Pearl Harbor. He concluded that the ash cones probably erupted during the latest ice age, but before about 30,000 years ago. K. A. Pankiwskyj concluded that the eruptions took place between about 400,000 and 100,000 years ago. That discrepancy leaves a wide spectrum of possibilities for future workers to resolve.

The Salt Lake cones are famous in the geological world for their xenoliths of pyroxenite and lherzolite, fragments of the earth's mantle as much as 6 inches across. Some lherzolite xenoliths contain red garnet, a mineral rare in Hawai'i. Laboratory work on experimental synthesis of minerals leads most geologists to believe that rocks of this type originate at depths of at least 30 miles.

The easiest place to see these ash deposits is in the large cut behind the baseball backstop in Salt Lake Crater District Park, at the end of Ala Lilikoi Place. This cut is in the ridgeline between the Salt Lake and Āliamanu ash cones; a country club occupies most of the wide, shallow Salt Lake crater. If you look north, you can see the rim of Āliamanu Crater, which is on a military base.

Pearl Harbor

West of Salt Lake Crater, the H-1 Freeway skirts the northern shores of Pearl Harbor. This famous anchorage is the composite estuary of several streams that joined at the harbor.

When sea level was low during the most recent ice age, old reefs, lagoons, and other sedimentary deposits were exposed in a broad coastal plain through which streams cut new valleys. As sea level rose at the end of the ice age, seawater flooded much of the coastal plain and the lower areas in the stream valleys and formed Pearl Harbor. Parts of the old divides between streams still stand above sea level, making up the peninsulas and islets that separate the branches of the harbor. The narrow entrance to Pearl Harbor is the flooded mouth of the main stream, which ran to the coast when sea level was low, cutting through the fringing reef to erode the passage used by ships today. If sea level remains stable for the next few thousand years, the streams that empty into Pearl Harbor will convert much of the anchorage into a coastal marsh.

The branches of Pearl Harbor are stream valleys that were flooded as sea level rose at the end of the latest ice age.

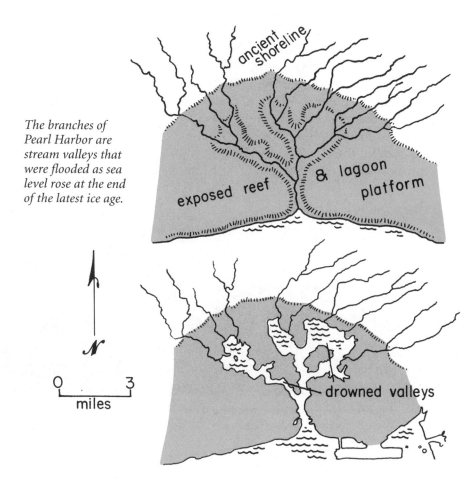

215

Across the Coastal Lowland to the Wai'anae Range

West of Pearl Harbor, H-1 heads toward the southern end of the Wai'anae Range. As much as a thousand feet of sediment deposited on hard basalt bedrock lies beneath the wide coastal lowland south of the freeway. This substantial accumulation includes sand and gravel laid down in streams and beaches, silt and clay deposited in coastal lagoons, reef rock, and even a few beds of coal. It was all laid down as the island slowly sank when it moved away from the Hawaiian hot spot.

Just west of Waipahu, the H-1 Freeway meets Kunia Road, Hawai'i 760 and 750. Kunia Road leads south 5 miles to 'Ewa Beach Park, crossing the coastal plain and passing through the town of 'Ewa.

Like almost all the beaches on O'ahu, 'Ewa Beach consists of soft, calcareous sand washed in from the fringing reef. This is a fine swimming area. Trade winds are continuous, and pleasantly warm.

Along the lower eastern flank of the Wai'anae Range, a series of alluvial fans coalesce into a broad plain. Streams flowing out of the mountains arrive at the Leilehua Plateau with a heavy load of sediment. They immediately dump most of that load onto the alluvial fans, partly because they slow down as they reach the gentler gradient of the Leilehua Plateau, and also because the water soaks into the extremely permeable sediments and the lava under it.

Watch along the slopes north of the freeway, between mileposts 4 and 3, for two eroded cinder cones well covered with vegetation. These erupted during the Kolekole phase of volcanism, the last gasp of Wai'anae Volcano. The small, grassy cone farther upslope is Pu'u Kapua'i, and the reddish hill directly ahead is Pu'u Makakilo. The H-1 Freeway skirts the southern base of Pu'u Makakilo. Roadcuts have exposed late alkalic lavas of the Pālehua member.

Kolekole cinder cone near the H-1 Freeway, in the southern Wai'anae Range. —David Saltzer photo

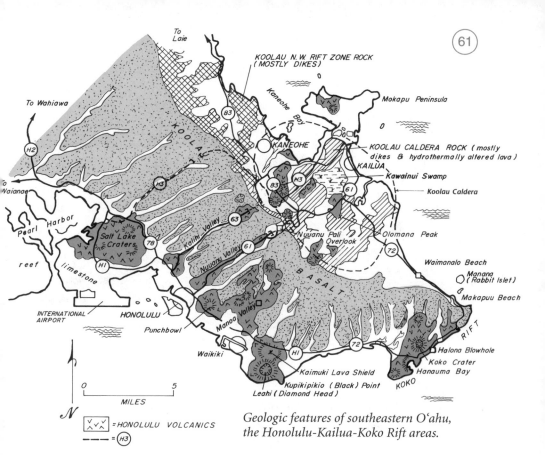

*Geologic features of southeastern Oʻahu,
the Honolulu-Kailua-Koko Rift areas.*

Hawaiʻi 61
Honolulu–Kailua
10 miles

A busy freeway, Hawaiʻi 61, winds through Nuʻuanu Valley, linking downtown Honolulu with Kailua-Kānéʻohe, one of the city's most important residential suburbs. It crosses the southern part of the Koʻolau Range and the eroded lowland of Koʻolau caldera.

Nuʻuanu Valley was deeper before recent eruptions during the Honolulu stage of volcanism partly filled it with lava flows of nephelinite, a variant of basalt so rich in sodium that it contains nepheline instead of plagioclase feldspar. These eruptions also built the flat valley floor that the highway follows as it climbs to Nuʻuanu Pass. Makīkī cinder cone, one of the volcanoes that erupted the nephelinite, appears about 5 miles up the slope from the junction with the H-1 Freeway.

Rocks exposed in the upper walls and head of Nuʻuanu Valley are the older thin basalt flows that built the giant shield of Koʻolau Volcano. The grade of the valley floor approximately matches the original slope of the shield.

The broad, eroded lowland of Koʻolau caldera spreads out below Nuʻuanu Pali.

Pali Lookout

Near the end of his campaign to conquer the Hawaiian Islands, Kamehameha the Great forced the army of a local chief up Nuʻuanu Valley, along the present route of Hawaiʻi 61. When they reached the head of the valley, the Oʻahu forces were trapped between Kamehameha's warriors and the Pali, a formidable cliff. They were unable to rally, and when Kamehameha drove three hundred of them off the cliff, the Oʻahu forces surrendered.

The Pali Lookout overlooks almost the entire amphitheater-shaped Koʻolau caldera. This huge, fluted cliff was left standing as less resistant rocks inside the caldera eroded. The low ridges in the caldera are stream divides that become much higher as they cross into the more resistant rocks outside the caldera.

Dikes that trend northwest to southeast in the Koʻolau rift system help support some of the high ridge crests outside the caldera. The dikes are impermeable and function as underground dams that trap groundwater. The reservoirs they impound are the major source of water for Honolulu.

A short walk southeast along the Old Pali Highway from the overlook leads past outcrops of dipping Koʻolau shield lavas, cut by dikes with glassy margins. About 700 feet from the overlook, the old road crosses a lava flow of dark nephelinite basalt that drapes the steep hillside. It erupted from a Honolulu-stage cinder cone perched on top of the ridge. With patience, you can locate a few weathered xenoliths of lherzolite in this flow, some as much as 4 inches across. They resemble the xenoliths in the ash that erupted at Salt Lake Crater.

A dike with tilted cordwood columns, Nuʻuanu Pali Overlook.

Kawainui Marsh and the Lowland Caldera

East of Nuʻuanu Pass, Hawaiʻi 61 quickly descends to the base of the Pali. Turn left onto Hawaiʻi 83, follow it for a mile, then turn east on the H-3 Freeway, toward Kāneʻohe Marine Corps Air Station. After about a mile, the H-3 Freeway passes a large rock aggregate quarry, to the left. Orange laterite in it caps pale gray lava flows.

About a mile beyond the quarry, you will see Kawainui Marsh, a large flat area to the south. Until geologically recent times, this region was a bay. Fine sediment that washed in from streams draining nearby Waimānalo Valley have almost completely filled the bay.

Hawaiʻi 61 meets Hawaiʻi 630 about 1.5 miles inside the eroded caldera of Koʻolau Volcano. On Hawaiʻi 61 east of the overpass, you will see some large roadcuts. They have uncovered a few of the many dikes that slice through the mass of rock fragments that were broken when the caldera collapsed and were altered by acidic vapors. At least five vertical dikes are exposed in the rusty roadcut.

Close inspection of the rock fragments shows that some of them contain pieces of older dikes, their vesicles filled with white silica, zeolites, and other minerals. Because vesicles can form only under very low pressures, these dikes must have intruded at a shallow depth. Caldera subsidence broke the dikes and surrounding rocks.

219

The nearly vertical strips in this roadcut are dikes cutting through brecciated lava flows inside Ko'olau caldera. —David Saltzer photo

After the rocks were shattered into a breccia, another generation of dikes, which appear in the roadcut, intruded it. These younger dikes lack vesicles, which suggests that they were emplaced under pressures too great to permit gas to bubble out of the magma. Evidently, a heavy load of lava flows already covered the area when they intruded.

According to geologist George Walker, the evidence provides a remarkably complete geologic history. First, the flows and early set of dikes helped to build the summit of Ko'olau Volcano. Then, when the caldera formed, the dikes and flows broke into rubbly fragments. As magma continued rising, dikes once again began to intrude the area, feeding lava flows that filled the caldera. Finally, erosion and highway construction stripped away the outer layers, exposing the deeper caldera structure.

Near the exit to Kāne'ohe Bay Road is another big roadcut showing many vertical dikes.

Hawai'i 72 to H-1
Kalaniana'ole Highway, around
the Southeastern Tip of O'ahu
Kailua–Honolulu
25 miles

Hawai'i 72 traces a route around the southeastern cape of O'ahu, passing young volcanoes of the Koko rift zone and interesting geological features along the shoreline.

The sharp peak south of Kailua is Olomana (1,643 feet), a stack of thick, horizontal lava flows that ponded inside Ko'olau caldera. Circulating hot water and steam altered the flows when the caldera was still active.

About a mile south of the junction with Hawai'i 61, watch for roadcuts where you can see weathered beds of broken rock fragments. This ancient rubble accumulated along the base of the fault scarps enclosing Ko'olau caldera. Erosion has obliterated the scarps.

Almost 6 miles southeast of Kailua, Hawai'i 72 approaches the sea cliff at the southeastern cape of O'ahu. Streams and waterfalls have eroded a striking pattern of channels in the cliff face. Many lava tubes, beheaded by erosion, have opened up high in the walls.

Waimānalo Beach, popular year-round, is a spectacular expanse of pale calcareous sand eroded from the wide offshore reef. Snorkelers can easily explore the reef top, which extends almost to the shore.

Two small islands appear offshore southeast of Waimānalo Beach. The larger and more distant is Mānana, or Rabbit Island from when a plantation manager used the island for raising rabbits; coincidentally, the shape of the

Salt weathering formed holes in the basalt south of Makapu'u Beach.

island suggests the head of a rabbit heading south, with its ears laid back. The other island is Kāohikaipu. Both are volcanic cones along the Koko rift zone, which trends southwest from here for about 7 miles. Sea Life Park, inshore from Rabbit Island, stands on the youngest flow of the series. Radiocarbon dates place the age of this pāhoehoe lava, a variety of basalt called basanite, at about 6,000 years.

The small rocky point on the coast between Rabbit Island and Sea Life Park is deeply pock-marked basalt, a vivid display of salt weathering.

Makapu'u Beach, south of Sea Life Park, is the state's most famous bodysurfing beach. With no offshore reef to block steep waves, they break hundreds of feet offshore and roll all the way in, over the gently sloping sandy bottom. Longshore currents feed two prominent rip currents. Much of the sandy beach is washed away by winter storms, revealing jagged rocks that the smaller waves of summer cover again with sand.

Nearby "Sandy" is one of the most popular hangouts for local bodysurfers. It is a beautiful stretch of pale calcareous sand similar to Waimānalo Beach. The sand offshore drops abruptly to a rocky bottom, creating a dangerous shore break. So many men drowned here during World War II that the military placed the area off limits.

The Koko Rift Zone

South of Makapu'u Point, the highway turns to follow the Koko rift zone toward Honolulu. This region is the scene of the most recent volcanic eruptions on O'ahu. A dozen large cones are left from those eruptions, in various states of erosional disrepair and exposure. Despite its name, the Koko rift zone

In the roadcut next to the Blowhole, you can see angular blocks of basalt and greenish gray ash that were blown out of Koko Crater.

Thin layers of ash in the flank of Koko Crater, southwest of the Blowhole, contain bits of white coral scattered through the ash beds in the cinder cone.

contains no visible rifts or fissures. The northeast to southwest trend does not match any older fault or other known geologic trend on the island.

Koko Crater, a cinder and ash cone about 1,200 feet high, is the highest of the Koko volcanoes. The road skirts the seaward flank, where you can see thin layers dipping seaward at moderate angles. The magma that erupted from Koko Crater met groundwater before it reached the surface, generating large volumes of steam. The resulting blasts were extremely violent, initially throwing ash and other pyroclastic debris over an area a couple of miles wide. As the eruption waned, ejected material simply piled up around the vent, building up the cone. Look for neatly layered deposits of ash between the pullout for the Blowhole and Hanauma Bay.

Most of the ash was blown horizontally across the land in clouds of hot steam and choking gas. As such clouds slow and begin to disperse, they drop their cargo of ash in neatly layered beds, which geologists call pyroclastic surge deposits. If you look carefully in the ash layers, you may see bits of white coral. These too must have been blown out in the ash explosions. Part of a reef or a pebbly coral beach apparently lies buried under the ash cone.

It is generally difficult to distinguish between airfall and surge layers in ash deposits. One clue is crossbeds, which form when surges of steam dump

Wave-eroded ash beds near Hālona Blowhole include both surge and airfall ash deposits. —Ronald K. Gebhardt photo

their ash in shifting beds that resemble sand dunes. Airfall ash deposits rarely contain crossbeds.

Watch for a pullout at the base of Koko Crater to view Hālona Blowhole. Waves rushing into a lava tube at the waterline compress air into the back of a cave. The air rushes out through a small opening in the roof, spraying water like a spouting whale.

Across the small inlet next to the Blowhole is a cliff face in the point of land where the rocks are mainly flat beds of volcanic ash. Scattered crossbeds indicate that eruptive surging played a role in depositing them.

Pellets of ash are abundant close to the pullout for the Hālona Blowhole. In a curious excess of terminology, the pellets smaller than peas are called pisolites, while the larger ones are called lapilli. Some geologists believe they form from clumping of ash by falling raindrops. Laboratory studies, however, show that they can also form as ash swirls through eruption clouds heavily charged with static electricity. The static charge clumps the particles of ash, the same way a charged comb attracts hair.

Layers of ash across the road from the overlook contain gently tilted crossbeds typical of surge deposits. Look for bits and pieces of dark gray basalt and white reef fragments embedded in the ash. The rush of erupting steam

224

Pisolites in ash at Hālona Blowhole. —David Saltzer photo

and gas blasted them out of the vent. Some of the reef fragments contain fossils. After erosion truncated the bedding across the top of the roadcut, debris washing down the slope covered the erosion surface.

The Lāna'i lookout parking area, southwest of Koko Crater, is a good place to see ash that erupted explosively from nearby Kahauloa Crater. The ash beds exposed in the roadcut have partially altered to yellowish palagonite. Look for crossbeds and fragments of reef limestone. Round pellets of ash abound shoreward of the parking area, and you may find raindrop impressions on the surfaces of some layers. Waves eroding the shore have exposed part of the ancient reef that was buried under the ash.

Pullouts near Hanauma Bay offer good views of remarkable exposures of surge deposits from nearby steam explosions. Look for ash layers sagging beneath the weight of blocks of basalt blown out after the ash.

Hanauma Bay is a flooded volcanic crater with a fringing reef that protects it from wave erosion. It is a marine sanctuary and a popular snorkeling area. A coral reef around the inner bay almost reaches shore. Heavy public misuse has killed most of the coral, but a visit to the bay is a must for thousands of tourists who want to see colorful fish up close.

The oldest radiocarbon age dates on volcanic rock from Hanauma Bay and Kahauloa Crater are about 32,000 years. Some ash layers contain blasted bits of reef rock that date as young as 7,000 years. These dates bracket the eruptions at between 32,000 and 7,000 years ago.

Exposed coral reef in Hanauma Bay.

The prominent bench at Hanauma Bay is being eroded by salt spray and salt weathering.

A prominent and interesting feature at Hanauma Bay is the flat ledge a few feet above water level that nearly encircles the bay. It almost looks like a road had been built around the shoreline. The ledge is above the reach of waves at high tide. Although it is not battered by surf, recent studies show that it is diminishing because of salt weathering and erosion from salt spray.

The Koko rift volcanoes built a peninsula that extends southwest of Hanauma Bay to Koko Head. The trend continues below sea level at least another 3 miles.

Hanauma Bay to Honolulu

Farther west, Kalaniana'ole Highway passes Wailupe Beach, where the shallow bottom is mud and jagged coral rock. A large reef fringes the coast about 1,500 feet offshore.

Nearby Kāhala Beach is 1,800 feet of imported calcareous sand next to a resort hotel. The sand was barged in from the island of Moloka'i.

The prominent low point on the western horizon, across Maunalua Bay, where the highway enters the eastern suburbs of Honolulu, is Kūpikipiki'ō, also known as Black Point. Beyond it rises the large ash cone of Lē'ahi, better known as Diamond Head.

Kūpikipiki'ō is a lava flow of alkalic basalt that erupted from the lower eastern flank of Diamond Head about a half million years ago and poured into the ocean. It stands high because the rock resists erosion. An affluent residential neighborhood now covers most of the flow. Walkways provide access to outcrops along the shoreline, but public parking is almost nonexistent.

Diamond Head

Diamond Head is certifiably free of diamonds. It acquired its name from calcite crystals, which nineteenth-century British sailors mistook for diamonds, creating a brief sensation. Hot water and steam circulating through the ash beds in the crater rim dissolved the calcite from fragments of reef limestone incorporated in the volcanic debris, then precipitated it to form the crystals.

Diamond Head reaches 761 feet above sea level. It is another ash cone that grew as rising magma encountered water at shallow depth. The series of steam explosions that followed ejected volcanic ash and cinders, large blocks of older volcanic rocks, and chunks of reef limestone. The volcanic debris reacted with hot water to make palagonite clay, which colors much of the volcano in shades of reddish brown. You can reach a street that leads inside the crater from the Monserrat Avenue–Diamond Head Road, which encircles the north side of the volcano. From a parking area inside the crater, a trail to the Amelia Earhart Monument high on the western flank of Diamond Head, 560 feet above, provides excellent views of this famous volcano that erupted so briefly. Take a flashlight for the long, dark tunnel on the trail.

Diamond Head Crater.

Volcanic Vents in Honolulu

Hawai'i 72 becomes the H-1 Freeway and enters central Honolulu from the east through the broad saddle between Diamond Head and the Ko'olau Range. The low hill in the saddle with a shallow crater at its summit is the Kaimukī lava shield, another souvenir of the Honolulu volcanic episode.

West of the Kaimukī lava shield, the freeway crosses the mouth of Mānoa Valley downslope from the University of Hawaii, Mānoa Campus. The campus is built on a flow of an extremely rare alkalic basalt: nephelinite that contains melilite. Minute white crystals of melilite are visible only with a microscope, and are not much to see even with that help. The flow erupted between 6,000 and 10,000 years ago from Pu'u Kākea (Sugarloaf), a cinder cone perched on the western rim of the valley. The lava flow and underlying sedimentary beds give the valley a wide, flat floor, ideal for the suburban developments that cover it.

Mānoa valley is slowly developing into an amphitheater-headed valley. These result when stream erosion combines with the continuous dissolving of rock by rainwater, which, on tropical mountainsides, is slightly acidic.

Pu'u 'Ualaka'a is another ash cone just west of Mānoa Valley and upslope from the H-1 highway. At 1,048 feet, it provides broad views over Diamond Head, Honolulu and Waikīkī, and Punchbowl Crater. You can reach it from

Wilder Avenue, which parallels the H-1 highway on the north. Follow Makīkī Street, then Round Top Drive to the state park. Roadcuts along the way have exposed thinly layered gray ash in the flank of the cone.

Punchbowl Crater dominates the foothill topography of Honolulu west of Mānoa Valley. Another large ash cone, similar to Diamond Head, it too is about a half million years old. The lavas that erupted from Punchbowl Crater contain xenoliths of peridotite and pyroxenite similar to those at nearby Salt Lake Crater, except they do not contain red garnet crystals. You can drive to the center of the crater, which is about a quarter mile across. It contains the National Memorial Cemetery of the Pacific.

Hawai'i 83 and 99/H-2
Kailua—the North Shore—across the Island to Pearl Harbor
60 miles

Hawai'i 83 heads north from Hawai'i 61 near Kāne'ohe on the windward coast, across the crest of the Ko'olau Range from Honolulu. It follows the rugged windward coast of northwest O'ahu, the island's North Shore, and loops around the north end of the Ko'olau Range to Hale'iwa. From there, Hawai'i 99 climbs south onto the gentle Leilehua Plateau at the center of O'ahu,

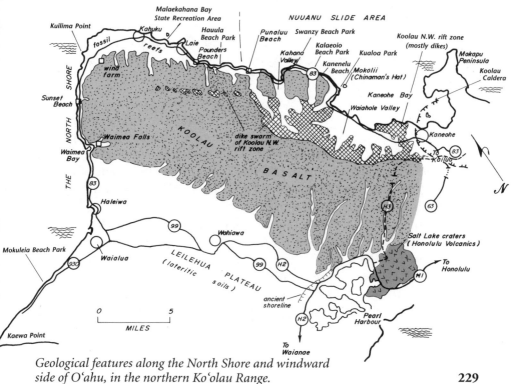

Geological features along the North Shore and windward side of O'ahu, in the northern Ko'olau Range.

229

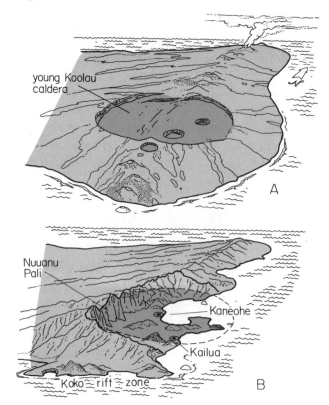

Collapse of the flank of Koʻolau caldera. A is before collapse; B is after collapse.

where it ends at Wahiawā. The H-2 Freeway continues south to join the H-1 near Pearl Harbor. The route provides excellent views of the shield volcano of Koʻolau, in varying degrees of erosional decay.

The Kâneʻohe Bay Area

The junction of Hawaiʻi 63 and 83, near Kāneʻohe, lies inside Koʻolau caldera, the collapse basin that lies above the extinct main magma chamber in Koʻolau Volcano. It forms the plain at the southern end of Kāneʻohe Bay and the adjacent lowlands around Kailua. The northwest rift zone of Koʻolau Volcano extends across Kāneʻohe Bay, parallel to the coast, indicating that the former axis of Koʻolau Volcano lies offshore in this area.

Near the southern end of Kāneʻohe Bay, north of Kāneʻohe, Hawaiʻi 83 climbs over the western edge of the caldera. Red laterite is exposed in a deep roadcut nearby. The fantastic fluted cliffs of Nuʻuanu Pali, reaching heights of 2,700 to 2,800 feet, loom immediately to the southwest.

Mokoliʻi Island is a large sea stack visible from the highway at the northern end of Kāneʻohe Bay, near milepost 30. Kualoa Park has good beaches and provides a convenient viewpoint. The island is an erosional remnant of a

Mokoliʻi Island (Chinaman's Hat).

resistant headland that once extended offshore. Waves tend to refract toward headlands, embracing them from both sides and concentrating their erosional energy, so they attack a headland more vigorously than a straight segment of coast. When erosion first separated the island from the mainland, sand eroded from the headland collected on the leeward side toward shore. A sandy peninsula, called a tombolo, probably connected the island with the mainland. With time, even the sand in deeper water washed away. The flat area of Kualoa Park is the remnant of that tombolo.

Waves attack solid bedrock mainly by compressing air into fractures in the rock, popping blocks of rock out into the surf. Ocean surf tends to exploit fracture zones, regardless of whether the rock is hard or soft. Most sea stacks, like Mokoliʻi, are remnants of former headlands that resisted erosion because their rocks are less fractured than others.

The cliffs of Moʻo Kapu o Hāloa Pali, across the north end of Kāneʻohe Bay, are about 1,900 feet high. They reach westward to the coast about a mile north of Kualoa Park. The prominent horizontal bands in the cliff face are thin shield-building lava flows from Koʻolau Volcano.

Kanenelu Beach, a little more than a mile north of Kualoa Park, is a steep, narrow strand of coral sand washed in from the reef offshore. The bottom near shore is sandy in some places, jagged reef rock in others, and the beach is popular with swimmers and surfers.

231

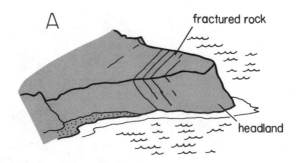

A

fractured rock

headland

B

C

sea stack

D

Formation of a sea stack.

Pali, cliff, above a patch of banana plants at Ka'a'awa.

Close by, Kalae'ō'io Beach lies inside a small bay, with reef rock extending all the way to the shore. One of the translations of the name Kalae'ō'io, "ghostly procession cape," associates this beach with spirits of the dead, who were said to enter the underworld through a lava tube at the foot of a nearby ridge spur.

Kāne'ohe Bay to Punulu'u—The Slide Coast

Along this stretch of coastline northward to Punalu'u, the highway skirts a steep mountain front that roughly follows the northwest rift zone of Ko'olau Volcano. This shore owes its steepness, and its proximity to the rift zone, to the enormous Nu'uanu Slide, which amputated a vast section of eastern O'ahu. Much of the eastern half of Ko'olau Volcano collapsed into the Hawaiian Deep after the rift was weakened by rising magma. Streams have dissected the slide scarp, and waves have eroded it to create the precipitous coastal mountain front.

The community of Ka'a'awa occupies the narrow wave-cut platform a little north of milepost 26. Cliffs a thousand feet high rise behind the town. Breaking waves mark the coral reef a couple of hundred yards offshore. The shallow water between the reef and the shore covers a somewhat lower wave-cut bench probably left behind as waves eroded the cliff landward.

In Ka'a'awa, Swanzy Beach is a narrow strip of calcareous sand at the base of a seawall. It was much larger until most of the sand was washed away in severe storms during the winter of 1968 and 1969. An offshore channel through the wide fringing reef carries a dangerous rip current.

The calcareous strand at Makaua Beach, a short distance to the north, is so narrow that it is completely underwater at high tide. Close to shore, the bottom is mainly very shallow reef, very sharp. A profile on the acute ridge protruding to the highway a half mile north of these beaches inspired its name, the Crouching Lion.

Kahana Valley State Park, at milepost 26, extends inland to include the huge, densely vegetated amphitheater at the head of the Kahana Valley. Its submerged extension, including Kahana Bay, was eroded when Oʻahu stood higher. Kahana is the largest drowned valley on the eastern shore.

Early efforts to secure a water supply for Honolulu focused on Kahana Valley. One drainage tunnel driven nearly 2,000 feet into the mountainside between 1929 and 1931 cut through 120 dikes in the northwest rift zone. The objective was to tap groundwater impounded behind the impermeable dikes, which act as natural underground dams. Tunnels such as this, in rift zones on Koʻolau, have become a major source of fresh water for the city of Honolulu.

Punaluʻu Beach is a stretch of calcareous sand immediately south of Punaluʻu. The beach is small, narrow, and subject to severe erosion, but it is excellent for swimming and snorkeling. The bottom is sandy, with patches of coral. The offshore reef is half a mile wide.

A stream enters the ocean at the northern end of Punaluʻu Beach. During the latest ice age, the stream eroded a valley to meet the shore about 2 miles east of the present coastline. Although this extension of the valley is now flooded, it remains as a deep channel with strong rip currents that slice across the offshore reef.

The Northeast Coastal Lowlands

Hauʻula Beach, at about mile 19, is narrow and about 1,000 feet long. The fringing reef directly offshore is about a quarter mile wide. Waves erode it to supply the beach with calcareous sand. The bottom is shallow and rocky along the central part of the beach. Deeper water off both ends of the beach is often rough.

About a mile south of Lāʻie, Hawaiʻi 83 skirts Lāʻie-Maloʻo Beach, which bodysurfers call Pounders because of the big shorebreak. This is a popular beach, a quarter mile of tan calcareous sand. The rip currents are ferocious.

A little over half a mile north of the Polynesian Cultural Center, Anemoku Street leads out to Lāʻie Point. A big roadcut along the street near the junction shows a section of beige fossil sand dunes that have cemented into solid limestone.

At Lāʻie Point, you can see old cemented sand dunes that once were heaped behind the beach and have eroded into coastal cliffs standing 20 to 30 feet above sea level. Just off the point, waves have eroded a low island out of the same material, forming a picturesque natural bridge. Strong waves are con-

Surf at Lāʻie Point is undercutting cemented fossil calcareous dunes. Waves have eroded the natural bridge on the island.

stantly undercutting the edge of the limestone, and blocks of the cliff occasionally topple into the surf. The thin, steeply dipping layers faintly visible in the cliffs were deposited on the downwind sides of the dunes, where sand accumulated in a cornice and slipped down the face.

A terrace of gravel and boulders on the plain southwest of the Mormon Temple in Lāʻie is the remnant of an ancient shoreline.

About a mile north of Lāʻie, look for a sign to Mālaekahana State Recreation Area, a small beach park with low overgrown sand dunes behind the sandy beach. Wave erosion of reefs offshore supplies the tan sand to Mālaekahana Beach, which extends for about a mile south.

The low offshore islets and the flat point at the northern end of the beach are remnants of an old reef now slightly above sea level. Rocks at the point include beautiful outcrops of fossilized calcareous algae, an important part of most reefs.

North of the old sugar mill at Kahuku, southeast of milepost 14, look for a hardened bed of pale boulders and patches of ancient reef rock seaward of the highway. This is an old shoreline about 25 feet above sea level. An old reef on the brushy slopes inland is about 100 feet above sea level. They apparently were formed when sea level stood high between ice ages.

A drop in sea level left this ancient fossil coral reef terrace at Kuilima Point.

About a mile east of Kawela, Kuilima Drive leads north from Hawai'i 83 to the Turtle Bay Resort at Kuilima Point. Kuilima Cove, at the eastern edge of the point, is the former mouth of 'Ō'io Stream, which has diverted to the east. A sheltering reef makes for calm water in the cove. Sand was imported to augment the beach.

When sea level was low during the most recent ice age, 'Ō'io Stream cut a channel through the offshore reef. These days, it draws a hazardous rip current along the west side of the cove.

Kalaeokamanu Point, on the eastern side of Kuilima Cove, is an old section of reef, standing a few feet above sea level. Deep holes in it, a few feet across, expose coral heads and thin platey algal layers.

Kawela Bay, at the eastern edge of the small town of Kawela, is one of the few places on the North Shore that is genuinely safe for swimming. The beach is several thousand feet long, with only a few rocky outcrops. Sand comes from reefs at both sides of the entrance to the bay. Even during heavy weather, the inshore swell is slight. The water is usually clear except in winter, when Kawela Stream empties sediment into the bay.

Not far away, Waiale'e Beach is long and wide, fine for beachcombing. Numerous outcrops of beach rock and coral in the shallow water make for difficult swimming. The fringing reefs that supply the tan sand extend several thousand feet offshore.

236

Ancient fossil reef and layered beach rock banked up on Pahipahiʻālua Beach, west of Kawela.

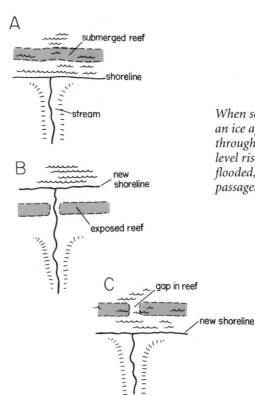

When sea level drops, as during an ice age, streams cut channels through exposed reefs. As sea level rises, the channels are flooded, creating deep-water passages that reach close to shore.

237

Concentric growth rings are visible in a fossil coral head about 3 feet across, at Pūpūkea Beach Park.

Coral pebbles at Pūpūkea Beach Park. View is about 4 inches across.

The North Shore of Oʻahu

Near here, Hawaiʻi 83 passes the northern end of the Koʻolau Range, where the northwestern rift zone of the old volcano meets the sea. Soil and vegetation cover most of the dikes. The coastline to the west features the most challenging surfing beaches on Oʻahu, and excellent examples of shoreline wave action. It is not unusual to see winter waves breaking above 20 feet here, driven by storms as far away as the Gulf of Alaska.

Kē Nui Road, a frontage road, begins in the town of Sunset Beach. Follow it one hundred yards west of Ekuhai Beach Park and you will reach an area where winter swells rise steeply as they climb a shallow reef about 150 feet offshore. Each monstrous wave breaks almost at once, curling from one end to the other into a tube known as the Banzai Pipeline. Surfers did not successfully ride these waves until the 1960s.

An ancient fossil reef offshore breaks the force of incoming waves at Pūpūkea.

Hawai'i 83 passes Pūpūkea Beach at mile 6.6, east of Waimea Bay. The shoreline is rocky, with little sand. Three Tables, at the west end of the beach park, and Shark's Cove at the east end are popular swimming areas. Some people regard Shark's Cove, with an assortment of tide pools, as one of the prettiest stretches of shoreline on the island. Concentric growth rings mark coral heads, several feet across, in the old reef. Pebbles of white coral are mixed in with the beach gravels.

A natural breakwater of ancient reef rock partly protects the inlet at Three Tables. The Three Tables are pieces of the reef with flat tops that stand above sea level.

Just 0.2 miles past Pūpūkea Beach Park, Pūpūkea Road leads upslope to Pu'u o Mahuka Heiau State Monument, overhanging the rim of Waimea Valley. The view is spectacular from the southern edge of the heiau, above the entrance to Waimea Valley. Waimea Bay occupies the flooded mouth of this sunken river valley. A sand bar and beach form the shoreline across the mouth of the valley. At low water, Waimea Stream cannot maintain a channel across the beach and merely soaks through the sand en route to the bay.

Waimea Valley is typical of hundreds of valleys and gulches along the northern and western flanks of Ko'olau Volcano. Almost nothing of the original surface of the shield volcano survives, but you can see evidence of the ancient slope in the general gradient of the divides between streams. Erosion has not

239

'A'ā flows exposed in section near the parking area at Waimea Falls Park.

progressed to the point of carving deep canyons or opening broad amphitheaters at the heads of the valleys.

Deposits of coarse gravel, boulders, and other sediments lie beneath the flat floor of lower Waimea Valley. They fill the floor of a canyon that was eroded when O'ahu stood higher above sea level.

Waimea Falls drops 80 feet across a resistant ledge of lava that crosses Waimea Valley. The falls may be reached at the end of a short walk through Waimea Falls Park. They are spectacular when the stream is full, but they dwindle to a trickle during dry weather. At the entrance to the park, outside the parking area, examine the cliff face to see the massive cores of 'a'ā lava flows. The lavas exposed in the Waimea Valley are basalt flows erupted during the shield-building stage of Ko'olau Volcano.

Waimea Stream is slowly filling Waimea Bay with sediment, patiently converting it into a marshy coastal flatland. Meanwhile, Waimea Bay is a world-class surfing locale. Although most of the giant breaks are out beyond the point, the inshore surf can break as high as 10 feet against the calcareous sand of Waimea Bay Beach. A strong rip current runs down the middle of the bay.

Roadcuts in the headland west of Waimea Bay have exposed several small basalt flows that erupted during the shield-building stage of Ko'olau Volcano. You can see pāhoehoe flows with lava tubes, and 'a'ā flows.

Kawailoa Beach stretches 5 miles along the coast between Waimea Bay and Haleʻiwa, with rocky outcrops sometimes interrupting the calcareous sand. During the winter, giant North Pacific swells move the sand offshore and shift the edge of the beach as much as 40 feet inland. Then the gentler waves of early summer bring the sand back.

Rocky Puaʻena Point separates Kawailoa Beach to the north from Haleʻiwa Beach to the south.

Across Leilehua Plateau

Hawaiʻi 83 leaves the north coast at Waialua Bay, a boat harbor at the north end of Haleʻiwa. Haleʻiwa Beach Park stretches along the eastern side of the bay. As you would suspect from the beige color, the fine calcareous sand was eroded from the reefs offshore.

Haleʻiwa Aliʻi Beach Park, at the western edge of the boat harbor, is a long stretch of beige sand with patches of old ragged reef terrace standing 5 to 10 feet above sea level.

South of Haleʻiwa and Thompson Corner, Hawaiʻi 99 and 803 both climb the broad slope onto the Leilehua Plateau. Fields of sugarcane, and later, as the grade eases, pineapples, extend for miles into the distance. Although pineapple is not the cash crop it used to be, the Dole Company's Helemano plantation covers a huge area.

Deep red lateritic soil, abundant in tropical climates, contrasts strikingly with the green sugarcane crops on the Leilehua Plateau. The laterite is more than 150 feet thick at the center of the plateau. This immensely thick layer

Sugarcane fields cover the plain below Mount Kaʻala. Deep gulches have been carved into the mountain's flanks.

of soil formed from sands and gravel that were washed off the adjacent mountain ranges, all within the past 2 million years.

Several deep gulches dissect the Koʻolau Range, in the distance to the east, but none have developed into broad amphitheater valleys as in the older Waiʻanae Range to the west. The contrast in rainfall may partly explain the difference, but the chief reason is that the Koʻolau Volcano is younger than the Waiʻanae Volcano.

A few miles south of Wahiawā and Schofield Barracks, a broad valley has eroded into the Waiʻanae Range to the west. The mountain with the flat top at the western end of this valley is Mount Kaʻala, the highest summit on Oʻahu (4,025 feet).

About 3 miles south of Wahiawā, near mile 6.5, the freeway crosses Kipapa Stream Gulch. Watch for beautiful deep exposures of the reddish brown laterite soil that covers the Leilehua Plateau.

Just north of Pearl Harbor, the freeway crosses a landscape of steep slopes and rugged gulches. This was the south shore of Oʻahu during a geologically recent high stand in sea level. Waves first shaped those slopes, then streams carved gulches into them after sea level dropped.

Fossil clam shells embedded in the concentric growth rings of coral heads in the fossil reef terrace at the end of the paved road in Kaʻena Point State Park.

Hawaiʻi 930, Westward to Road's End
Northwest Shore – Kaʻena Point
9 miles one-way

At Weed Traffic Circle, continue southwest along Hawaiʻi 930 to Thompson Corner, the junction with Hawaiʻi 803. Turn west along Hawaiʻi 930 to cross a flat coastal plain planted in sugarcane. The Waiʻanae Range, to the south, slopes steeply westward from Mount Kaʻala.

Mokulēʻia Beach Park, 5.4 miles west of Thompson Corner, is a pleasant beige sandy beach with layered beach rock lapping up at the water line. Breakers crash on the coral reef 100 to 200 yards offshore.

The brushy ridge a few hundred yards inland exposes several lava flows in the upper part of the Waiʻanae Range, the older volcano that makes up the western part of Oʻahu.

The road continues west along sandy beaches at the base of the north end of the Waiʻanae Range. The pavement and road end abruptly 8.4 miles west of Thompson Corner. An old reef is exposed in an ancient wave-cut bench that stands about 10 feet above the water. Waves break close to the shore here because the modern reef lies just offshore. It is fun to walk across the old reef, spotting concentric growth rings of coral heads, some clusters of tiny coral with radiating ribs, net patterns of bryozoa, and clam shells embedded in the coral.

A dirt track leads to the Natural Area Reserve at the rocky limits of Kaʻena Point, an additional 3 miles. It follows a nearly flat, raised terrace 100 to 200 yards wide. Reefs are at the shoreline most of the way.

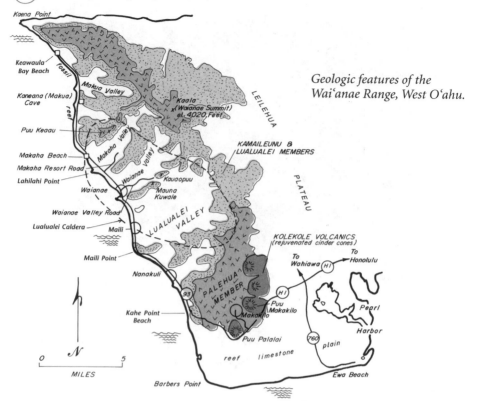

Geologic features of the Wai'anae Range, West O'ahu.

The Leeward Coast
Hawai'i 93, Farrington Highway
H-1 Freeway–Yokohama Beach
20 miles

From the junction with H-2 at Waipahu, H-1 crosses the south end of the low Leilehua Plateau. Deep red laterite soil contrasts with the green vegetation. The dry, leeward side of the island is much less cloudy than the windward side, and is mainly open range land.

Not far west of Pu'u Makakilo, the H-1 Freeway narrows and becomes route 93, Farrington Highway, which follows the west coast of O'ahu. It skirts the eroded caldera of Wai'anae, and ends in the volcano's eroded northwest rift zone near the island's northwest cape.

The Leeward Shore South of the Caldera

Just north of the junction with the H-1 Freeway, Farrington Highway turns north along the leeward shore, passing towering slopes eroded into many thin lava flows. Sea cliffs near the electric generating plant expose basalt flows that dip seaward, away from the southern rift zone of Wai'anae, which erupted the lava. They are part of the original shield volcano.

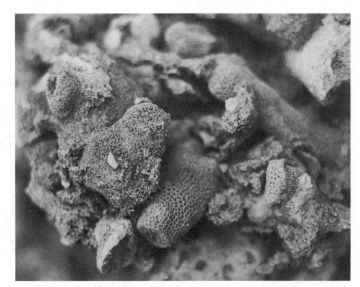

Coral in the fossil reef at Kahe Point Beach Park.

Despite its name, Kahe Point Beach Park contains very little beach. A low sea cliff that fronts the shore is part of an ancient reef that grew when sea level stood slightly higher than it does today. If the waves are low, you can walk down to the reef to see where it grew on outcrops of black basalt, which show evidence of salt weathering. Similar reef rock a few feet above sea level encircles most of the island. Take time to explore the old reef for fossil coral, bryozoa, algae, mussels, and snails. Hardy swimmers can enter the water via a small cove at the east end of the park.

Lualualei Caldera and the Nearby Coast

Lualualei, the wide valley floor that opens inland between Nānākuli and Haleakalā avenues, is the eroded caldera of Wai'anae Volcano. It is between 4 and 5 miles across and 11 miles long, and is rimmed with towering cliffs or steep hills. You can drive a short way into the caldera on the public part of Lualualei Naval Road, Hawai'i 780. The low, brownish white cuts along this road are limestone reef deposits that grew when sea level was slightly higher than now, during some earlier interglacial time.

Ulehawa Beach Park stretches north of Piliokahe Beach Park directly offshore from Lualualei Valley. It contains no substantial beach, just 2 miles or so of rocks, part of an ancient reef terrace.

At the north end of Ulehawa Beach is Mā'ili Point, a bold, dark red prominence directly north of Lualualei Valley. It is a remnant of the old Wai'anae Volcanic Series flanking the Lualualei caldera. Lava flows from it dip seaward away from the center of the caldera.

Farrington Highway passes Mā'ili Beach Park less than a mile north of Mā'ili. The calcareous beach sand tends to wash away during winter storms,

At the entrance to Lualualei Valley, lava flows can be seen in the cliff where it slopes away from the eroded caldera of Wai'anae Volcano.
—David Saltzer photo

Pu'u o Hulu Uka, an erosional remnant of the Wai'anae shield, forms Mā'ili Point. —David Saltzer photo

Looking toward the head of Lualualei Valley from Wai'anae Valley Road. The foreground ridge to the right crests at Mauna Kūwale, a rhyodacite mass. A steplike cut for a water tank terminates the ridge. Low to the left is Kauoapuu Ridge, which is also capped by rhyodacite lava. —David Saltzer photo

then returns in the late spring. In many places, it is overgrown with grass. At the north end of Mā'ili Beach Park is another dark red, rocky knob, part of the southwest wall of the caldera.

Immediately adjoining Mā'ili Beach to the north is Lualualei Beach Park, which continues into Wai'anae. The shoreline consists of rugged beach rock with layers that dip seaward, as the old beach did.

Just north of Wai'anae, the highway meets the Wai'anae Valley Road. Outcrops near the head of Wai'anae Valley expose a type of rock that is rare on oceanic islands, a high-silica lava called rhyodacite. Private land and dense vegetation prevent easy access to the best exposures. Rhyodacite ordinarily erupts in continental volcanic fields. Why is it here? Some geologists believe that it is residue from crystallization of a basaltic magma chamber that stagnated for a long time toward the end of Wai'anae shield growth.

Lahilahi Point

Lahilahi Point between Wai'anae and Mākaha is a small, steep-sided peak, Mauna Lahilahi. Separating two small bays, both with fine beaches, it is an erosional remnant of basalt shaped like a giant blade 230 feet high. The southern beach is Mauna Lahilahi Beach Park. Calcareous sand washes onto the beach with the gentle waves of summer and is carried away with the heavy surf of winter. A low, wave-cut bench around the point is cut into basalt flows and capped by white fossil reef and layered beach rock. Reef features are fairly well preserved, including some staghorn coral. You can reach Papaoneone

247

Mauna Kūwale. —Ronald K. Gebhardt photo

Staghorn coral in the fossil reef at Lahilahi Point.

Mauna Lahilahi, an isolated remnant of basalt, forms Lahilahi Point, at the north end of Lahilahi Beach. —David Saltzer photo

Beach on the north side of the peak via Moua Street. The water deepens near shore, with a strong longshore current.

Across Farrington Highway from Mauna Lahilahi Beach Park, Mākaha Valley Road leads inland to a resort and country club. The towering cliff face of Pu'u Kea'au, 2,650 feet high, rises north of the golf course. Several thin lava flows stand out in it. Those nearest the ocean are thin and dip seaward, but the ones inland are thick and nearly horizontal. If you look carefully, you will see a fault separating them.

Harold Stearns concluded years ago that the fault was the rim of Lualualei caldera. This site may record the filling and overflowing of the Wai'anae caldera near the end of volcanic activity. If so, the inland flows would indeed be horizontal and thick because they ponded inside the caldera, and the seaward flows would have overflowed the rim of the caldera to pour down the flank of Wai'anae shield volcano.

Mākaha Beach Park, near the mouth of Mākaha Valley, is one of the world's classiest surfing locales. When the surf is high, strong longshore currents merge into a dangerous rip tide at about the middle of the beach. This current flows out to sea along a former channel of Waiele Stream, which cut across the beach when sea level was lower.

North of Mākaha, the highway passes Kea'au Beach Park, about 2 miles of ancient reef terrace. Dangerous offshore currents run along this section of coast.

North of the Caldera

Farther north, Farrington Highway threads its way between the ocean and the steep western ridges of the Wai'anae Range. These ridges are in the northwest rift zone of the old shield volcano. Many basalt dikes slice through the dark flows of basalt that built the main part of the volcano.

Puʻu Keaʻau from Mākaha Valley. The fault that defines the caldera margin is directly above the tree at lower right. Horizontal ponded flows are to the right; seaward-dipping flows outside the caldera are to the left.

Entrance to Mākua Cave. Vertical dikes in the cliff are part of the Waiʻanae northwest rift zone.
—Ronald K. Gebhardt photo

The tops of some of the ridges carry a veneer of pale alkalic basalts that erupted during the later stages of volcanic activity. Like the older flows, these dip gently seaward.

While large portions of the western flank of the Waiʻanae Volcano slid into the ocean, this part did not collapse catastrophically. Sonar surveys of the ocean floor show that the few large fragments are still nearby, indicating that they slid piecemeal in a series of slow slumps.

Mākua (Kāneana) Cave, the Cave of the God Kāne, is south of the mouth of Mākua Valley. Waves carved out this enormous sea cave when sea level stood almost a hundred feet higher along this shore; geologists refer to that era as the Kaʻena high stand of sea level. Ocean waves exploited a swarm of fractures to open a passage about 450 feet long and 15 to 20 feet high. The coastal platform shoreward of Mākua Cave is a reef terrace that records another high stand in sea level, later and lower than the Kaʻena stand.

The many lighter gray dikes that make vertical stripes 2 to 5 feet wide in the nearby sea cliffs show that this area lies in a rift zone. If you look carefully, you will see that the edges of the dikes have darker border zones. Rising basalt magma cooled against the older rock to become finer grained and darker. This is the northwest rift zone of Waiʻanae Volcano, which extends nearly 70 miles from the site of Lualualei caldera to the deep ocean floor, just northwest of Kauaʻi. Only about 15 miles of it are above sea level.

Wave-eroded ancient reef terrace near the north end of Farrington Highway, south of Kaʻena Point. —David Saltzer photo

Look inland a mile or so north of Mākua Cave to see a beautiful landscape with a broad amphitheater at the head of the valley. This is Mākua Valley. Several basalt dikes of the northwest rift zone cut through the southern slopes. A long beach of beige reef sand stretches between rocky points at the mouth of the valley.

The pavement ends at Keawaʻula Bay Beach, sometimes called Yokohama Bay because of its popularity with Japanese fishermen. The dirt road beyond clings precariously to a narrow fringe along the base of the mountain front and is no place to go unless you have a four-wheel-drive vehicle. The track leads to the Kaʻena Point Natural Area Reserve.

The Keawaʻula shoreline is rocky. At the north end of the beach, you can walk along the crest of an old raised reef terrace, the flats next to the shore. Look for several species of coral, along with mussel and periwinkle shells, especially in reef cliffs in the small protected bays. The partly cemented beach and dune sands covering sections of the terrace must have accumulated when sea level stood slightly higher than now. Look for the crossbeds that formed as waves shaped and reshaped the ancient beach.

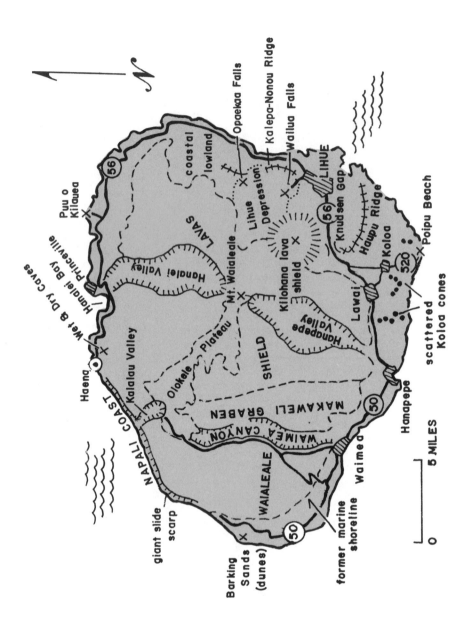

Major geologic features of Kaua'i.

7
Kaua'i, The Garden Isle

Kaua'i is the oldest, most deeply eroded, and most heavily vegetated of the Islands. Many people consider it the loveliest. The top of a single enormous shield volcano, Wai'ale'ale, its summit is at 5,170 feet. Kaua'i is roughly 25 miles long, 33 miles across, and has an area of 555 square miles. Despite appearing to be geologically simple, the rocks and terrain of Kaua'i tell a complex and unusual story.

Growth of Wai'ale'ale

Wai'ale'ale began to build above sea level sometime between 5 and 6 million years ago. Eventually, it grew into a gigantic shield volcano, like Mauna Loa on the Big Island. Thin flows of pāhoehoe and 'a'ā lava, each from 5 to 15 feet thick, piled up to shape the volcano's gently sloping flanks. Eruptive fissures radiated from the summit toward all points of the compass, though most were concentrated in two rift zones trending west and northwest.

Pioneer Hawaiian geologists Harold Stearns and Gordon Macdonald concluded that Wai'ale'ale developed an uncommonly large summit caldera about 13 miles across. They named it Olokele caldera. They based their thinking mainly on a massive sequence of horizontal lava flows about 2,600 feet thick on the Olokele Plateau, at the center of the island. Such a thick stack of flows must surely have ponded in some sort of basin, which Stearns and Macdonald believed could only have been a giant caldera. They interpreted a similar stack of flows in Hā'upu Ridge as the remains of a smaller caldera, which you can see in Queen Victoria's Profile, south of Līhu'e.

The idea that the Olokele Plateau and Hā'upu Ridge are remnant calderas poses problems because both of them are resistant highlands. All the other extinct calderas in the Hawaiian Islands have eroded into broad natural amphitheaters at the heads of valleys.

Now that sonar surveys have given them detailed topographic maps of the ocean floor, some geologists see the history of Kaua'i differently. Several fields of rubble on the ocean floor appear to be massive pieces that broke off the eastern part of the island and slid into the ocean. One mass moved north, the other south. Debris from them swept up the slope of the Hawaiian Arch, as far as 60 miles from Kaua'i. Slides with such tremendous momentum must have been moving very fast and surely raised big tsunami.

The slide scarps split the summit of Wai'ale'ale along a line a short distance east of Waimea Canyon, roughly parallel to it. What remained of the eastern part of the island had dropped. A strip of the volcano about 3 miles wide and 15 miles long dropped along parallel faults to become the Makaweli graben, which extends from the northwestern part of the island to the southern shore.

Another catastrophic slide later broke a large chunk off the north flank of Wai'ale'ale and carried it into the ocean. A towering sea cliff was left behind, which has since eroded into the Nā Pali Coast.

Wai'ale'ale attempted to repair itself: Eruptions continued after the slides, building a new volcanic shield in the dropped eastern region of the island. Lava streamed down the new shield to fill the Makaweli graben, and it ponded against the towering slide scarps that faced east in the Olokele area, building the thick stack of massive flows in the Olokele Plateau. Lava also ponded against irregular slide and slump blocks near Līhu'e, forming the great stack in Hā'upu Ridge.

The new shield developed a caldera, which ultimately eroded into the Līhu'e Depression, a broad basin 5 miles northwest of Līhu'e. Gravity surveys have revealed that the earth's gravitational field is slightly stronger than normal above the floor of the Līhu'e Depression. A large mass of dense rock probably exists beneath the floor of the depression. Many geologists think it is a mass of gabbro, the crystallized magma chamber of Wai'ale'ale.

Geologists call the Wai'ale'ale shield lavas the Waimea Canyon basalt. One part is the Nā Pali member, which includes the thin lava flows on the slopes of the Wai'ale'ale volcanic shield. Another part is the Okokele member, which includes the thick lavas of the Olokele Plateau. Similarly thick flows that went through and filled the Makaweli graben are called the Makaweli member.

Rapid shield building, and rebuilding, continued at Wai'ale'ale until 4.3 million years ago. As activity waned, the lavas became more alkalic and the eruptions more explosive. This late activity was under way by 3.9 million years ago, and added little to the island. Meanwhile, erosion had already begun to slowly dismantle Kaua'i.

Early Erosion

We can imagine those first stages of erosion: Gigantic slices of the volcano continued to slide away, extending the coastline seaward in some places, shifting it inland in others; waves eroded the shorelines into sea cliffs, except where coral reefs protected the coast; and streams dissected the originally gentle volcanic slopes into a landscape of jagged ridges and deep valleys.

The sloping flanks of Wai'ale'ale eroded far more rapidly than the ponded lavas in the Olokele Plateau and Hā'upu Ridge. The flank flows were thin, inclined seaward, and full of weak layers of permeable rubble. Streams easily

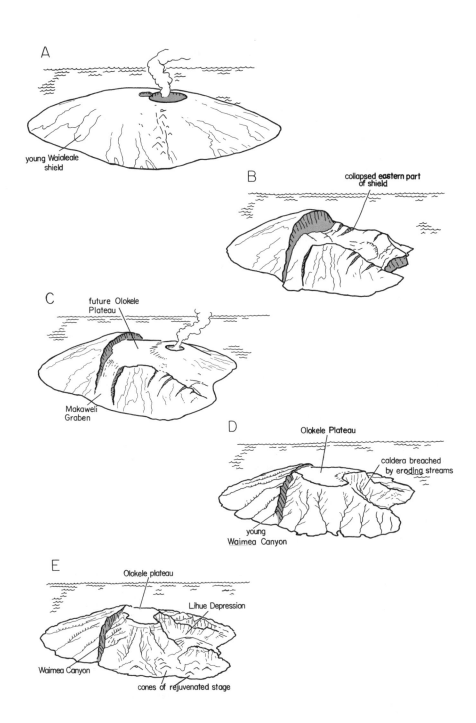

A

young Waialeale
shield

B

collapsed eastern part
of shield

C

future Olokele
Plateau

Makaweli
Graben

D

Olokele Plateau

caldera breached
by eroding streams

young
Waimea Canyon

E

Olokele plateau

Lihue Depression

Waimea Canyon

cones of rejuvenated stage

Stages in the development of Kaua'i.

undercut them along the rubble layers, triggering slides that followed the tilt of the flows. Only a few striplike remnants of the original flanks of the shield volcano survive.

Massive stacks of thick Olokele and Hāʻupu basalt flows, which originally ponded in low areas, today rise above the surrounding landscape, standing in bold erosional relief. The network of deep stream valleys that dissects the Kauaʻi highlands still does not completely drain the Olokele Plateau, where some of the heavy rainfall, instead of running off, collects in high swamplands. The swamps support a lush montane forest that shelters some of the rarest bird and plant species in the Islands.

Volcanic Rejuvenation

Kauaʻi would be smaller than it is and even more rugged were it not for an episode of renewed volcanism that took place mainly between 3.65 and 0.5 million years ago. As on east Oʻahu, these youngest eruptions were generally small and widely scattered. Geologists call the rocks produced by these eruptions the Kōloa basalts.

About forty late volcanic vents have been identified on Kauaʻi. The oldest are in the west and northwest parts of the island, the younger ones in the east and southeast. The largest is Kilohana, the gently sloping mountain immediately west of Līhuʻe. A small shield volcano 6 miles across, it has a summit crater about 250 feet deep. Most of the vents from the rejuvenated stage line up along north-to-south trends, with a lesser number lying along northeast-to-southwest trends.

The Kōloa basalts poured into many valleys, completely filling some of them and diverting their streams to erode new channels elsewhere. Much of the landscape of eastern Kauaʻi expresses that long interplay of streams and eruptions. Lava erupting next to the shore added new land to Kauaʻi, including much of the island's coastal flatlands.

As on Oʻahu, the alkalic basalts from the rejuvenated volcanic stage include some exotic rocks greatly enriched in sodium, such as nephelinite, melilite basalt, and basanite. Most of the flows contain fragments of lherzolite and dunite. Most, if not all, of the xenoliths are bits of the mantle beneath Kauaʻi, a part of the earth we would know nothing about without them. Geologists love them.

Late Erosion

In the past half million years or so, streams have deeply eroded the Kauaʻi valleys that escaped the effects of renewed volcanism, and have eroded new valleys in the Kōloa volcanic fields.

In Hanalei Valley, near Princeville, streams have cut through a section of Kōloa lavas more than 2,000 feet thick. Hanalei Valley extends from the coast

12 miles inland to the summit of Waiʻaleʻale, and is as much as 4 miles wide and 3,000 feet deep, with nearly vertical walls. The neighboring valleys to the west, Lumahaʻi and Wainiha, are equally spectacular, as is Waimea Canyon on the opposite side of the Olokele Plateau. Together they testify to the power of weathering and erosion in a warm, wet climate.

Some 5 million years ago, when Kauaʻi first emerged from the ocean, it was a steamy volcanic summit perched directly over the Hawaiian hot spot, where the Big Island is now. Five million years from now, the ocean may reclaim it as the plate beneath cools and subsides during the journey to the Kurile trench. By then, Kauaʻi will probably be a mere coral atoll awash at high tide. Enjoy it while it lasts.

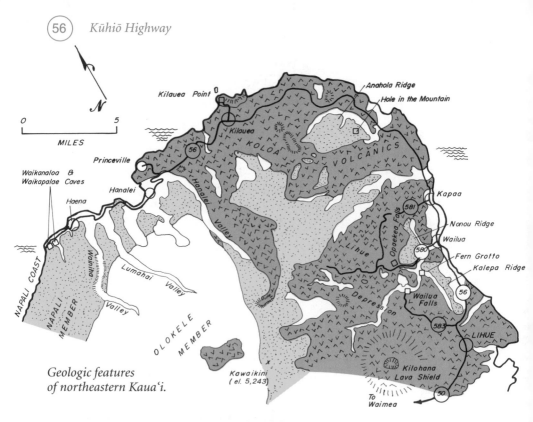

Geologic features
of northeastern Kaua'i.

Wailua Falls.

The North Shore
Hawai'i 56, Kūhiō Highway
Līhu'e–Hā'ena
38 miles

The Kūhiō Highway skirts the east and north coasts of Kaua'i. For about 10 miles north of Līhu'e, it follows a low coastal route, close to easily accessible beaches. The beige beach sand consists mainly of fragments eroded from the offshore reefs. Coconut palms and other tropical vegetation thrive in the warm, moist climate. Young sedimentary deposits in the river valleys and beneath the coastal lowland are intensely farmed, except along the highway. There they support an exuberant growth of motels, shopping centers, and housing developments.

Farther north and west, sea cliffs that were eroded in Kōloa basalts force the road to turn inland for a mile or so to the smoother upland surface. There it passes through gently rolling countryside, past orchards, fields of sugarcane, and undeveloped brushy acreage. Lush tropical forests grow in the stream valleys that cut through the uplands.

West of the resort community of Princeville, the road again hugs the coastline, passing close to lovely beige sand beaches and climbing over rocky headlands. Paved side roads provide access to rocks that record most stages of the island's dramatic volcanic growth and slow erosional decay.

In the first mile north of Līhu'e, the highway crosses a tract of old sand dunes, now for the most part overgrown. They probably record a time when sea level stood lower, perhaps during the latest ice age, which ended about 11,000 years ago. After the reefs and lagoons were left high and dry above the new shoreline, wind blew the sand inland.

You can see evidence of a lower sea level on Kaua'i because it is one of the older islands, well beyond the hot spot and now sinking very slowly. You do not expect to see evidence of lower sea levels along the shorelines of young Hawaiian islands, because they are sinking as they grow larger and heavier.

Wailua Falls Area

About a mile north of Līhu'e, route 583 branches to the northwest, into the southern part of the Līhu'e Depression, the eroded remains of the only known caldera on Wai'ale'ale. A short way from Hawai'i 56, route 583 passes a roadcut that shows an excellent example of horizontally layered volcanic rocks being weathered into rounded residual stones.

Wailua Falls, 3 miles down the road, is a double cascade about 80 feet high in the south fork of the Wailua River. At the falls, the stream is eroding into beds of ash and weathered pillow lava beneath a thick, resistant lava flow of Kōloa basalt. Slabs occasionally break off the undercut lava flow, keeping the lip of the falls sharp.

261

The basalt at the rim of the falls is a pāhoehoe lava flow that poured into a muddy swamp or deep pool on the floor of the ancient Wailua River valley. The flow is thick because it was confined between the valley walls. The lower 10 to 30 feet developed into oversize lava pillows as the lava encountered standing water. Some of the pillows are 15 feet across, giants of their kind. The upper part of the flow is ordinary pāhoehoe lava, with no pillows. Later eruptions nearly filled the valley with more Kōloa flows and ash. The modern stream is now busily eroding all of these young rocks out of the valley.

Southwest of Wailua Falls, you can see the gently sloping profile of Kilohana shield volcano. Directly west is the edge of the Olokele Plateau and Kawaikini, the summit knob of Waiʻaleʻale, which is 5,243 feet high. To the north, you see ʻAʻahoaka Hill, an erosional remnant of Nā Pali lavas surrounded by younger Kōloa flows. Look northeast to see the notch that the Wailua River has cut through Kālepa-Nounou Ridge; it too is composed of Nā Pali flows.

North to Fern Grotto

A mile or two north of the junction with route 583 on Hawaiʻi 56, look west to see Kālepa Ridge. It is a remnant of older Nā Pali lavas that stood high while younger Kōloa lavas flowed around it. The central part of the Līhuʻe Depression is just beyond Kālepa Ridge.

About half a mile north of milepost 5, Leho Road leads to the remains of Puʻuhonua o Hauola, an ancient Hawaiian place of refuge and redemption for people who had violated kāpu (taboos). About 0.3 miles farther north, Hawaiʻi 56 reaches the junction to Wailua River Marina, where the famous boat tour to the Fern Grotto begins. The Wailua River carved the cave from a weak flow of Nā Pali basalt covered with comparably weak ancient stream sediment. The Fern Grotto roof is a resistant flow of Kōloa lava.

ʻŌpaekaʻa Falls

At milepost 6, route 580 leads inland 3 miles through Wailua River State Park to ʻŌpaekaʻa Falls. The Kauaʻi Historical Society and Bishop Museum have restored the ancient temple of Holoholoku Heiau near the highway intersection. Human sacrifices were offered at the stone in the southwest corner of the heiau.

Route 580 passes through the notch that the Wailua River eroded across Kālepa-Nounou Ridge. Kōloa lavas poured through the notch, so it must have been in existence before the stage of rejuvenated volcanic activity began. The river is now eroding these young flows.

ʻŌpaekaʻa Falls is a pretty little waterfall that tumbles across Kōloa basalt. Just below the falls, you can see vertical dikes of basalt cutting through horizontal flows of Kōloa lava. The dikes fill fissures, at least a few of which fed lava into the uppermost flows.

Cliffs of Nā Pali basalt rise above the coastal plain near Anahola, inland from Hawai'i 56.

Rocks on the slopes east of 'Ōpaeka'a Falls are shield volcano lava flows, part of the Nā Pali basalts. Some of the many dikes that cut across these older flows along Kālepa-Nounou Ridge dip toward the center of the Līhu'e Depression. Harold Stearns and Gordon Macdonald interpreted them as typical cone sheets, curving dikes shaped like segments of the cones that generally enclose calderas. Cone sheets are extremely rare, and typically are found in more deeply eroded volcanic regions, such as Scotland, where they were first recognized. If these are indeed cone sheets, they are the only ones identified in the Hawaiian Islands.

Wailua to Anahola

Hawai'i 56 crosses old beach deposits between Wailua and the village of Kapa'a. If you look inland, you can see the northern continuation of Kālepa-Nounou Ridge. The sharp silhouette along the skyline outlines the profile of the Sleeping Giant: The head lies to the south, the feet to the north. The generally gradual descent of the main ridgeline toward the coast preserves one of the few remnants of the original slope of Wai'ale'ale. Almost everywhere else on the island, the slope has been lost to erosion.

Watch for the scenic lookout about a mile north of Kapa'a, half a mile from milepost 9. Immediately north and just below the lookout is a wave-cut

Wave erosion has exposed ash layers in the side of a Kōloa cone across the small bay from Kīlauea Point.

bench of basalt, a relic from a time when sea level was slightly higher. A mile farther north, the highway follows a long beach of tan calcareous sand. Like most beaches on older Hawaiian coasts, this one was formed as waves eroded the offshore reefs, washing the sandy debris ashore.

Between mileposts 13 and 14, the highway passes through the small community of Anahola. Look inland from milepost 14 to see a jagged ridge, another eroded remnant of the original shield volcano. Hole-in-the-Mountain is a small natural arch near the ridgeline, where a landslide sheared the surface off the thin crest of the ridge, exposing an old lava tube.

North of Anahola, the road crosses an eroded coastal platform made of poorly exposed Kōloa basalt flows. Puʻu ʻAuʻau, the small hill north of the road at milepost 17, is a small lava shield and cone, one of the younger Kōloa vents.

Puʻu Kīlauea

Puʻu Kīlauea is the prominent hill projecting from the shore about 1.5 miles north of milepost 22. It erupted sometime between 13,000 and 15,000 years ago, in what was probably the most recent volcanic activity on Kauaʻi. It is an important seabird sanctuary, a nesting ground for the red-footed booby. About 0.4 mile north of milepost 23, at the community of Kīlauea, a side road leads across 2 miles of coastal tableland to Puʻu Kīlauea and the sanctuary.

The eruption of Puʻu Kīlauea began when rising magma met groundwater at a shallow depth, flashing it into steam in a series of violent explosions. The cone is a disorderly mixture of volcanic ash, fragments of white limestone from the reef, and blocks of older Kōloa lava that the steam explosions ripped out of the vent. The blocks include chunks of rare melilite basalt and nephelinite. The ejecta accumulating around the vent during the final phase of the eruption cut off the magma from the ocean, bringing an end to the steam blasts. The remaining molten rock simply fountained out of the ground, spreading a coat of spatter across much of the cone.

For a good view of that volcanic potpourri, look across the small bay east of Kīlauea Point. The cone is on the coast, about half a mile away. The tiny islet off the end of the point is Moku ʻAeʻae, part of the far side of the cone. A sand bar connects the islet to the shore at low tide.

At the western edge of the path to Kīlauea Point, look down to see a black lava flow with well-defined columns.

Kalihiwai Bay

Hawaiʻi 56 meets Kalihiwai Road about a mile west of the Kīlauea Point Road. The village of Kalihiwai is 1.1 miles down Kalihiwai Road. The tsunami of April 1, 1946, heavily damaged Kalihiwai and destroyed an important bridge, separating the east and west ends of Kalihiwai Road. You will have a good view of the coral reef extending offshore from the beach cliff as you approach Kalihiwai Bay. The line of breakers marks the offshore edge of the reef.

A vertical basalt dike about 15 feet wide is visible in the cliffs at the east end of Kalihiwai Beach. The fissure it fills was probably the plumbing that fed molten lava into one of the Kōloa basalt flows. The dike shows prominent horizontal columns, which are basically like the vertical columns in lava flows. Both form at right angles to the cooling surfaces as molten lava shrinks while it crystallizes into solid rock.

Fragments of the volcano's mantle— peridotite xenoliths in nephelinite boulders at the eastern end of Kalihiwai Beach.

265

The black boulders at the eastern end of the beach look like ordinary basalt, but they are nephelinite, a rare rock type so rich in sodium that it contains abundant crystals of nepheline rather than ordinary plagioclase. The boulders eroded from a nearby Kōloa lava flow.

Some of the boulders contain fragments of peridotite, xenoliths an inch or two across, which appear as angular pieces even darker than the enclosing basalt. A determined search will also turn up a few xenoliths of dunite, a yellowish green rock consisting almost entirely of olivine. Olivine weathers readily as its iron reacts with atmospheric oxygen, which explains why some of the xenoliths have the rusty color of iron oxide. All of the xenoliths are fragments of the mantle, direct evidence of where the nephelinite magma originated.

'Anini Beach to Hanalei Valley Area

The lookout on Hawai'i 56 next to the Kalihiwai River, west of milepost 25, provides a picturesque view of Hawai'i as it would be without so many visitors. The river meanders quietly through small, fertile fields. Upstream, a waterfall pours onto the west bank.

'Anini Road branches from the highway between mileposts 25 and 26 and leads to 'Anini Beach Park, which many connoisseurs consider the finest snorkeling locale on Kaua'i. The long, exposed offshore reef protects the shallow inshore water from high surf, except during rough winter storms.

Hanalei Valley.

The sandbar at the mouth of a stream near the western end of the beach is a good place to look for seashells.

Kōloa lavas built the broad, flat peninsula at Princeville, the site of an old Russian fort and today the site of a modern resort. West of Princeville, Hawaiʻi 56 degenerates into Hawaiʻi 560, much of which is narrow and winding, with many one-way bridges. Mileposts begin again from zero.

The highway west of Princeville follows the breathtaking North Shore, the most deeply eroded flank of Waiʻaleʻale volcano. Watch 0.2 mile west of Princeville Center for a scenic lookout where the road begins to descend into Hanalei Valley. The view inland up the length of the valley reveals a lush tropical paradise. Fields of kalo (taro), a Hawaiian dietary staple, cover the floodplain of the Hanalei River below. The low ground about 1.5 miles upriver is in the Hanalei National Wildlife Refuge.

The massive ridge along the east side of Hanalei Valley was once a canyon almost 2,000 feet deep. Then thick flows of Kōloa lava filled it to the brim. The lava displaced the canyon's stream to the west, where it eroded thin and erosionally weak Nā Pali basalt flows to a depth of about 1,000 feet, creating the modern Hanalei Valley. The eastern ridgeline is a mold of the ancient canyon cast in basalt.

The highway curves through a big switchback, the Loop, as it enters Hanalei Valley. Farther west, the highway winds past the mouth of Lumahaʻi Valley, then Wainiha Valley. On a clear day, you can see the rim of the Olokele Plateau on the skyline at the head of the Wainiha Valley.

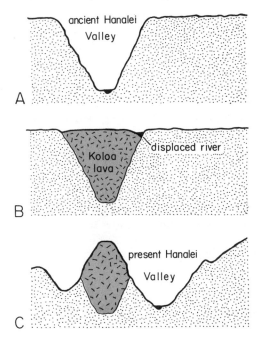

Displacement of Hanalei Valley by Kōloa lava flows.

Lumahaʻi Beach. Sand collected in coves like this eroded from basalt headlands and from coral reefs offshore.

The old whaling port of Hanalei is at the mouth of Hanalei Valley, between mileposts 2 and 3. You can charter a boat to explore the spectacular Nā Pali coast, which stretches 5 to 15 miles westward. Hanalei Beach is the wide strip of calcareous sand west of the pier. Swimming here can be rough.

Watch east of milepost 5 for the pulloffs to Lumahaʻi Beach (Waikoko Beach); this was the setting for the movie *South Pacific*. The trail to the beach begins at a prominent bend in the road.

Lumahaʻi Beach contains many grains of transparent olivine in beautiful shades of yellowish green. If you have a magnifying glass or hand lens, you can also see pretty little bits of coral, calcareous algae, and the minute shells of the order Foraminifera, tiny animals that drift in the ocean.

The scattered outcrops of basalt sticking through the beach contain depressions filled with sand extremely rich in olivine. Waves swashing sand around in the holes have washed the lighter mineral grains away, leaving behind dense concentrations of the heavier olivine, like miniature placer deposits. The largest outcrop is an offshore rock that would be worth climbing for the view were it not for the dangerous surf.

Just west of milepost 5, look for big roadcuts in neatly layered, gently dipping beds of volcanic ash weathered to a reddish brown. Two basalt dikes, each

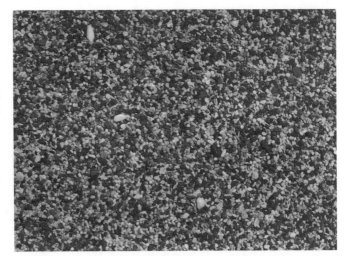

Sand on Lumahaʻi Beach is rich in glassy green olivine and white coral, algae, and tiny shells.

about 2 feet across, cut steeply through the ash. The fissures they filled were probably the magma plumbing for lava flows that have been lost to erosion.

Hāʻena and Road's End

Just west of milepost 8, the road passes through Hāʻena, which was devastated in the great tsunami of 1946. The highest wave crested here at 45 feet.

Nearby, cars are often parked around a short, inconspicuous road that leads to Tunnels Beach. Patches of coral reef almost reach the sandy shore, marine life is abundant, and the water is calm. The outer of the two parallel reefs offshore catches most of the surf.

The Kōloa basalts end near Hāʻena. Rocks farther west are older Nā Pali basalts that erupted during the main shield-building stage of volcanic activity.

About a mile west of Hāʻena, the road reaches Hāʻena Beach Park. Snorkeling is usually rewarding on the reefs here, where convict fish congregate to feed on the coral.

Watch south of the road for Maniniholo Dry Cave. It has a wide entrance and a broad ceiling, but not much depth. The floor typically is dry, which is why people call it a "dry cave." Waves eroded Maniniholo from flows of blocky ʻaʻā and ropy pāhoehoe basalt before the beach grew large enough to protect the sea cliff. The lavas erupted during the early shield-building stage of volcanism. The steep dikes cutting through them are in the northwest rift zone.

Hawaiian tradition holds that the fire goddess Pele created Waikapalaʻe Cave during her search for a new home. It is 1 mile west of Maniniholo. The floor of the cave extends below the water table, so it is flooded, and therefore called a "wet cave." Waves eroded Waikapalaʻe into horizontal lava flows, so it is surprising that the mouth of the cave is round rather than elliptical.

269

Maniniholo Dry Cave.

Waikapalaʻe Wet Cave.

Looking west from Hāʻena along the precipitous Nā Pali Coast on Kauaʻi. The old reef is partly exposed along the beach in the foreground. Kēʻē Reef is visible offshore.

The road ends at Kēʻē Beach in Hāʻena State Park. A reef with a conspicuously flat top lies a short distance offshore, protecting the small curve of a beach. Knobs of reef rock, built from coral and calcareous algae, rise vertically from the ocean floor, nearly reaching the surface. Beautiful ripple marks streak the sandy bed between the reef and the shore. Snorkelers can see a good variety of tropical fish and some colorful patches of growing coral.

A trail leads from the end of the road above the rocks to a former Hawaiian hula temple, where young people learned ancient mele (chants) and other traditions of their culture.

The Kalalau Trail also starts at the end of the road, winding westward for about 11 miles along the remote Nā Pali Coast. The terrain is rugged and dangerous along much of the distance, especially beyond the first 2 miles. A few hundred yards up the trail, openings through the trees provide fantastic views over Kēʻē Beach and the coral reef protecting it from the big offshore waves. Shades of turquoise tell the water depth: pale in shallow water, darker in deeper water.

Hanakāpīʻai Beach, a mile or so west, is bouldery, which probably reflects the scale of the surf that built it. The heavy seas also eroded the cave in the sea cliff at the west end of the beach. A trail leads from east of Hanakāpīʻai Beach to a high waterfall about 1.5 miles inland.

Kēʻē Reef viewed from the Kalalau Trail.

Hanakāpīʻai Beach, along the Kalalau Trail. The boulders were left behind after heavy storms.

Most of the remainder of the trail follows the contours of very steep slopes along the cliffy shore. About 10 miles from the end of the road, the trail crosses the lower end of Kalalau Valley, which you can also see from the high lookouts near the end of the Waimea Canyon Road. The trail ends at Kalalau Beach, where waves have eroded two caves into the sea cliff at the western end.

The Nā Pali Coast

Whether you see it on foot, from roadside lookouts, from the ocean, or from the air, the uninhabited northwest coast of Kaua'i is one of the most dramatic meetings of land and water anywhere. Geologists believe the steep, plunging cliffs were established several million years ago, when a large piece of the original north side of Kaua'i broke off and slid into the ocean, leaving behind a cliff as much as 2,700 feet high at the headwall of the slide. This was when the volcano Wai'ale'ale was entering its dotage of alkalic eruptions. Erosion has cut deep valleys and canyons into the old sea cliff.

Interior slopes along the Nā Pali Coast are sharply steep because of the thick lava flows of the Olokele Plateau, which is only 3 miles inland. They enable the top of the plateau to stand high because they resist erosion far more effectively than the surrounding Nā Pali flows, which degrade into gentler slopes.

Dike with steeply inclined shrinkage columns, along the Kalalau Trail.

Thin flows of Nā Pali basalt lie beneath the slopes of the narrow ridges that descend like long tendrils from the Olokele Plateau to the ocean. Large dikes radiating from the old caldera support most of the narrow ridge crests. Like the plateau lavas, they do not erode easily and persist as high ridges while the less resistant flows of Nā Pali basalt erode.

Although the landscape of the Nā Pali Coast seems eternal in the context of a human life span, it is in fact extremely unstable, rapidly changing in the longer spans of geologic time. You may witness small changes in occasional commonplace events such as rockfalls or when muddy streams carry sediment to the ocean. Multiply those by millions, and the result is a landscape in rapid transition.

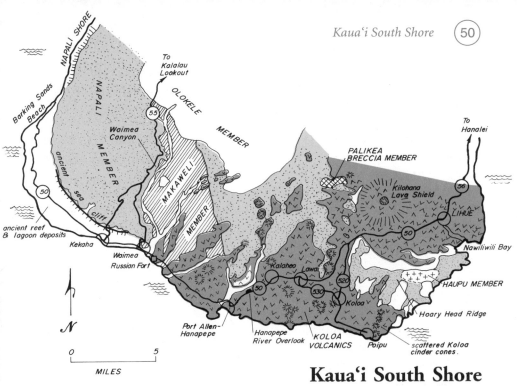

To
Kalalau
Lookout

Barking Sands
Beach

NAPALI SHORE

NAPALI MEMBER

Waimea
Canyon

OLOKELE MEMBER

MAKAWELI MEMBER

ancient sea cliff

ancient reef
& lagoon deposits

Kekaha

Waimea
Russian Fort

To
Hanalei

PALIKEA
BRECCIA MEMBER

Kilohana
Lava Shield

LIHUE

Nawiliwili Bay

Kalaheo   Lawai

HAUPU MEMBER

Hoary Head Ridge

Koloa

N

0   5
MILES

Port Allen-
Hanapepe

Hanapepe
River Overlook

KOLOA
VOLCANICS

Poipu

scattered Koloa
cinder cones

Kaua'i South Shore
Hawai'i 50
Līhu'e–Barking Sands
36 miles

The skyline south of Hawai'i 50 in the area west of Līhu'e is a ridge with a flat top rimmed with cliffs. People call it Hā'upu Ridge, Hoary Head, or Queen Victoria's Profile. Rocks on the ridge are a thick stack of horizontal lava flows that probably ponded against the headwall of a slide. Their excellent resistance to erosion explains the height of the ridge. To the northwest along the same stretch of road, you see the low profile of Kilohana, a shield volcano 1,133 feet high, the largest of the Kōloa volcanoes.

Less than half a mile west of Līhu'e, Hawai'i 50 meets route 58, which leads 2 miles to Nāwiliwili Bay, the main port on Kaua'i and the site of a large bulk sugar plant. The rapid rise in sea level as the ice melted at the end of the latest ice age, approximately 11,000 years ago, flooded the mouth of a valley to create this bay.

All of the bedrock in the Nāwiliwili Bay area is part of the younger Kōloa volcanic series. These flows dip southward, away from the Līhu'e Depression from which they may have erupted. Flows in nearby Hā'upu Ridge are nearly horizontal.

Watch for the deep roadcut on Hawai'i 50 about 0.7 mile west of milepost 5. Deeply weathered lava flows of Nā Pali basalt are visible in it, with basalt dikes cutting steeply through them.

At the western end of Hā'upu Ridge, Hawai'i 50 crosses Knudsen Gap, a low pass. This is a broad remnant of a deep valley eroded through lava flows of Nā Pali basalt. Younger flows erupted from neighboring Kilohana and other Kōloa volcanoes partially filled it, which explains the level floor.

'Ōma'o is west of Knudsen Gap, between mileposts 8 and 9. Look south to see Manuhonohono, a lava shield built of Kōloa basalt flows that erupted along an older rift zone in Wai'ale'ale.

At Lāwa'i, about half a mile past milepost 10, Hawai'i 50 crosses the eastern flank of a large gulch eroded in flows of Nā Pali basalt. Roadcuts west of the gulch reveal flows of younger Kōloa basalt with layers of pebbly conglomerate sandwiched between them. The thin layer of red at the top of each conglomerate bed is an ancient soil baked underneath the lava flow.

The Kōloa volcanic pile contains many buried red soils. They show that Kōloa volcanism continued intermittently over a long period, with enough time between eruptions for soil to develop. The conglomerate beds suggest that some of the Kōloa flows blocked stream drainages, causing floods to dump coarse rubble behind the lava barriers rather then carrying it to the coast. The loose flood rubble later hardened into solid conglomerate.

Kalāheo is slightly less than a mile north of Kukui o Lono, a prominent Kōloa cone. Several north-to-south alignments of Kōloa lava shields and cinder cones are in this area, but they are difficult to see from the road.

The highway crosses a flat coastal lowland between Kalāheo and Hanapēpē Bay. This shelf is composed of Kōloa lava flows, ash beds, and related sedimentary deposits, in many places weathered to red laterite.

Look to the north about 0.3 mile west of milepost 12, at the junction with Hawai'i 540, Halewili Road. You can see another cluster of small Kōloa volcanoes, the Pohakea Hills, about 1.5 miles away.

Hanapēpē Valley Area

East of Hanapēpē, the highway skirts the lower reaches of rugged Hanapēpē Valley. The lookout north of the highway, west of milepost 14, offers a spectacular view.

The thin flows in the higher valley walls are Nā Pali basalt that erupted on the flank of Wai'ale'ale. After Hanapēpē Stream eroded the valley into these shield lavas, thick flows of Kōloa basalt partly filled it. During long interludes between eruptions, Hanapēpē Stream eroded the flows, only to have its work undone when new eruptions again poured molten basalt into the valley. This was repeated several times during the period of rejuvenated volcanism. One big flow with conspicuous vertical columns stands out on the eastern side of the canyon.

The floor of Hanapēpē Valley becomes wide and flat as it approaches the coast. The stream eroded the valley when Kaua'i stood higher than it does

Pillow lava in the wall above Menehune Ditch, near Waimea.

today. As the island sank, the mouth of the valley became Hanapēpē Bay. Sediment deposited in the bay built the low plain where Port Allen stands.

Hawai'i 50 meets Hawai'i 543 at the western edge of Hanapēpē. This road leads south for about a mile to some ancient salt ponds next to the old airstrip at Pū'olo Point. The ponds were built to extract salt from seawater. Clay from the red laterite soil was used to seal them.

Salt Pond Beach Park, about half a mile west, is a sandy beach sheltered behind small coral reefs at the entrance to Hanapēpē. Coral grows right up to the shore.

Coastal Flatlands around Waimea and Kekaha

West of Salt Pond Beach, Hawai'i 50 crosses more coastal flatland. Look north at Kaumakani to see Pu'u o Pāpa'i, another small Kōloa volcano. About 0.6 mile west of Kaumakani, next to the underpass, a roadcut exposes red laterite buried under an eroded Kōloa basalt flow.

The remains of a Russian fort built in 1817 are in Elizabeth State Park, on Lā'au'ōkala Point.

Captain Cook landed at Waimea, at the mouth of Waimea Canyon, in 1778. Menehune Road, near the Captain Cook Monument in the center of Waimea, leads about 2 miles north to Kīkī a Ola (the Menehune Ditch), a moat under the end of a suspension footbridge. The earliest Hawaiian im-

An ancient shoreline inland from Hawai'i 50, west of Waimea.

migrants built this ditch centuries ago. Carefully shaped stone blocks and a tunnel testify to their skill and industry.

A layer of pillow lava that flowed out of nearby Waimea Canyon is visible in the cliff above Kīkī a Ola. The pillows range from 1 to 10 feet in diameter and show crude radial cracks that opened as the crystallizing lava shrank. Glassy bits of weathered lava, now turned to mudstone, partly fill the small spaces between pillows. The vertical shafts in the cliff face are molds of burned tree trunks. Lava pillows together with tree molds like this are highly unusual. Perhaps the lava flowed into a forested marsh on a river floodplain, allowing pillows to form around tree trunks and downed logs.

An Emergent Sea Floor

West of Waimea, Hawai'i 50 passes through Kekaha and crosses a dry plain, a bit of old sea floor with sand dunes on it. The plain originated when sea level was higher than today, presumably when the global climate was warmer and land ice melted. Waves then passed across the platform and eroded the low cliff at the foot of the slope north of the highway.

The beige soil on the plain resulted from the weathering of reef sand and silt that covered the platform. Some of that sediment settled underwater. The rest blew in from the beach during the latest ice age, when sea level was low and the dunes were moving. Low ridges near the coastal edge of the plain are old beach ridges and sand dunes.

The small streams emptying onto the plain from the flank of Wai'ale'ale have been channeled into a long ditch that is used for irrigating sugarcane fields. It skirts the hillside from Waimea west to Mānā.

Kekaha Beach Park, the waterfront at the western edge of Kekaha, has a beautiful beige sand beach—grains of coral and calcified algae washed in from

Polihale Beach, with the westernmost cliff of the Nā Pali Coast in the background.

offshore reefs. It extends for several miles along the southwest coast of the island. The surf is gentle, but the currents here are sometimes dangerous.

Look southwest from Kekaha to see the profile of Niʻihau, the westernmost of the main Hawaiian Islands. Inland is the gently sloping flank of Waiʻaleʻale, the island's big shield volcano. Streams have not dissected this dry side of the island as much as they have the wet side, and long strips of the original surface of the shield volcano survive intact on the drainage divides.

Barking Sands and Polihale Beach

Just west of milepost 32, Hawaiʻi 50 meets the road to Barking Sands Missile Range and Airfield, which occupies 7 or 8 miles of the coastal zone. The road jogs inland before continuing north parallel to the coast. Follow the signs to Polihale State Park.

Barking Sands is an area of large coastal dunes of beige calcareous sand. If the sand is properly damp, it really will "bark" as you walk across it. Like dune fields elsewhere in Hawaiʻi, these grew most rapidly during the ice ages, when the low sea level exposed reefs and offshore sediments to the wind.

The coastal road north of Barking Sands ends at Polihale State Park. Tracts of sand dunes covered mainly with small trees and brush back the long, spectacular beach. The beige sand is bits of coral and calcareous algae peppered with grains of black basalt eroded from the sea cliffs to the north. Waves wash the sand onto the upper beach at high tide. Then, as the sand dries at low tide, the sea breeze blows it off the beach and into the dune fields.

A walk to the far northern end of the beach leads to spectacular sea cliffs of Nā Pali basalt. Ocean waves are still eroding this part of the cliff. The dry coastal platform protects it farther south. Lava flows exposed in the cliffs slope gently seaward, parallel to the preserved remnants of the original flank of the volcanic shield above the cliffs.

Hawai'i 520 and 530, Kōloa Road
Hawai'i 50–Po'ipū Beach
5 miles

Kōloa Road is a north-south drive between rows of tall eucalyptus trees, called Eucalyptus Avenue, or the Tunnel of Trees. The long hill east of the road was eroded in the older flows of Nā Pali basalt before the Kōloa lavas flowed around them.

Kōloa, about 3 miles from Hawai'i 50, is a nicely restored old plantation town with false front buildings that date from the nineteenth century. The first sugar mill in Hawai'i was built here in 1835. The road jogs west in Kōloa, then again heads south as route 530.

Manuhonohono Hill, a prominent lava shield a mile west of Kōloa, is one of a series of rejuvenated-stage volcanic vents aligned along a north-to-south trend.

The big sand dunes southeast of Kōloa probably date from the latest ice age, when sea level was more than 300 feet below its present height. Now partially cemented into limestone, they lie on the Kōloa volcanic rocks that crop out between the dunes and the shore.

Route 530 reaches the south coast about 2 miles south of Kōloa. To the west, in Prince Kūhiō Park, is Hō'ai Heiau, an ancient temple and the birth-place of Prince Jonah Kūhiō Kalaniana'ole, a descendant of the kings of Kaua'i who for many years was the Hawai'i delegate to Congress.

Spouting Horn.

Poʻipū Beach is protected by an offshore coral reef.

Geometrically regular shrinkage fractures radiate through basalt on the shore at Poʻipū.

A side road leads to Spouting Horn, on the coast about 2 miles west of the highway. Waves sweep into the open mouth of a lava tube, sending fountains of water shooting up through a hole in the roof. When sea level stood about 5 feet higher, waves eroded the flat wave-cut bench through which the water spouts. Tan beach rock is exposed at the heads of small coves.

East of the intersection where route 530 reaches the coast, Poʻipū Road stays a bit inland, but another road follows the beach to Poʻipū Beach Park. This sunny, crescent beach is sheltered behind an outcrop of basalt that was a small island until waves built a sand spit tying it to the shore. Hurricanes battered the island ferociously in 1982 and 1992, and the damaged coral reef will take years to recover.

Cones and Dunes Near Poʻipū

Less than a mile northeast of Poʻipū Beach, north of Poʻipū Road, is Puʻu Wanawana, a symmetrical cinder cone breached on one side. The breach opened when a big piece of the cone floated away on a lava flow erupting from its base.

A neighboring vent, Puʻu Hunihuni, is less than 1.5 miles northeast of Poʻipū Beach. It is a lava shield that grew during one of the most recent Kōloa eruptions, less than 20,000 years ago. Puʻu Hunihuni may be as young or possibly even younger than Puʻu Kīlauea, on the North Shore. Two less prominent lava shields grew north and south of Puʻu Hunihuni during the same eruption. The southernmost of these three vents stands just east of Poʻipū.

A group of sand dunes cemented into solid limestone is northeast of the lighthouse in Poʻipū, at Makahūʻena Point. They were probably actively moving until about 11,000 years ago, when the great glaciers melted at the close of the latest ice age and sea level rose rapidly. Some of the dunes were submerged, along with most of their sand supply.

To the west, Hawaiʻi 530 links Kōloa to Lāwaʻi. This part of the route skirts the northern flank of Manuhonohono, one of several small lava shields aligned along a north-south trend that may be an extinct rift zone of Waiʻaleʻale Volcano.

Puʻu Wanawana, a breached Kōloa cinder cone near Poʻipū.

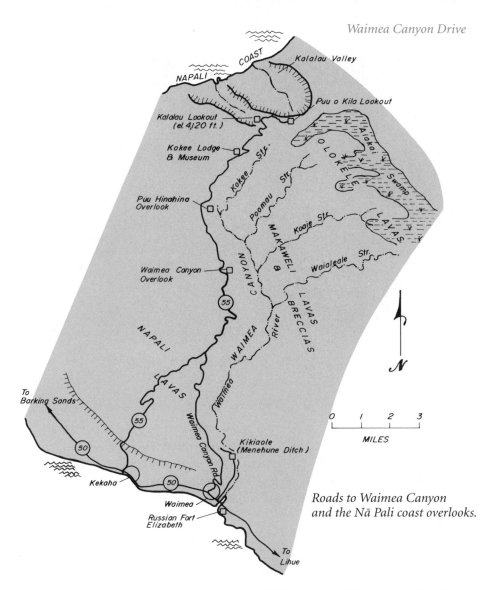

Roads to Waimea Canyon and the Nā Pali coast overlooks.

Waimea Canyon Drive
Hawai'i 50–Hawai'i 55
6.3 miles

Waimea Canyon Drive branches off Hawai'i 50 on the western side of Waimea, just west of milepost 23. The road climbs steeply up an ancient sea cliff, to a more gentle slope that is a remnant of the original flank of Wai'ale'ale. There it joins Hawai'i 55. A large pulloff about 0.6 mile past milepost 3 provides a good view—in clear weather—north up Waimea Canyon. The many lava flows in the walls of the canyon are Nā Pali basalt. They dip parallel to the old flank of the volcano.

283

Weathered flows of pāhoehoe lava along Waimea Canyon Drive, near milepost 6. Dark iron oxide stains at the base of the slope highlight the lava lobes.

Severe soil erosion and gullying between mileposts 5 and 6 illustrate the consequences of vegetation destruction in this wet climate. A roadcut about 10 feet high at milepost 6 reveals pāhoehoe flows that derive their red color from intense weathering. The red rings outline individual lobes of lava exposed in cross section.

Hawai'i 55
Kekaha–Waimea Canyon and Kalalau, Pu'u o Kila Lookouts
19 miles

This route leads to the fantastically rugged interior of Kaua'i, including the Olokele Plateau. It also permits the only roadside views from the crest of the precipitous Nā Pali Coast.

Along most of this route, the highway skirts the edge of panoramic Waimea Canyon, which is 14 miles long, as much as 2,500 feet deep, and exposes much of the interior structure of Wai'ale'ale. The remarkably beautiful canyon, with

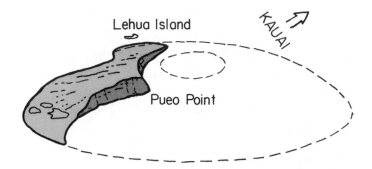

Lehua Island

KAUAI

Pueo Point

Niʻihau was a vastly larger island before most of it slid into the ocean.

red weathered rocks and lush green vegetation, is a paradise for photographers—especially those smart or lucky enough to arrive early in the day, before the clouds roll in.

Island of Niʻihau

The lower 5 or 6 miles of Hawaiʻi 55 provide a fine view southwest to Niʻihau, the westernmost inhabited island of the main chain. You can also see its small neighbor, Lehua, a partially submerged late-stage tuff cone. The eastern flank of Niʻihau is a towering cliff, the scarp at the headwall of a slide that detached a large section of the island and dumped it into the ocean hundreds of thousands of years ago. Only a small remnant of the original western flank remains above sea level.

A vertical dike 12 feet wide is exposed in the road bank near the intersection of Waimea Canyon Drive and Hawaiʻi 55.

Shells of weathered rock enclose residual stones in a dike near the intersection of Waimea Canyon Drive and Hawai'i 55.

Waimea Canyon

About 7 miles upslope from Kekaha, Hawai'i 55 intersects Waimea Canyon Drive. Some 150 feet beyond, look for a vertical dike about a foot wide that cuts through horizontal lava flows. It shows the well-defined shells of weathered rock enclosing residual stones about a foot across. Not far beyond this point, the road reaches the rim of Waimea Canyon.

Waimea Canyon is one of the world's truly stunning landscapes—intensely multicolored, eroded into weird pinnacles and spires, laced with waterfalls, and often shrouded in mists.

Big stream canyons ordinarily branch into smaller valleys upstream, with tributaries entering from both sides. Not so in Waimea Canyon. Here, the tributary valleys enter only from the east. Half of Waimea Canyon's tributaries seem to be missing.

A map of Kaua'i helps solve the mystery: Waimea Canyon cuts across the regional stream pattern. The strange tributary system developed because rain falling east of the canyon drains toward it, whereas almost all rain falling to the west drains away, down the western flank of the island. Why did the canyon develop in this orientation? Waimea Canyon follows the western edge of the Makaweli graben, the trough that dropped between faults and then was filled with lava when Wai'ale'ale was still an active shield volcano. Evidently, the Waimea River originally flowed to the sea along the graben. Lava from later eruptions displaced the river a few miles to the west, where it carved the present canyon. So the ancient slide scarps guided the erosion that shaped the modern landscape.

The upper part of Waimea Canyon roughly follows the eastern edge of the Olokele Plateau. The flat horizon of the plateau, seen across Waimea

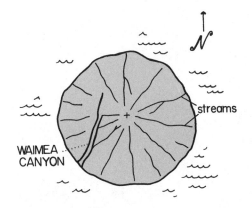

streams

WAIMEA
CANYON

A simplified map of Kaua'i, showing the radial pattern of streams flowing from the high center of the island and Waimea Canyon cutting across the western side of the wagon-wheel pattern.

Canyon, preserves a bit of the original lay of the land as it was before streams dissected Wai'ale'ale.

Within 2 miles of the entrance to Waimea Canyon State Park, look for small outcrops of gabbro along the road. This badly weathered rock consists of large crystals of dark pyroxene, white plagioclase, and occasional grains of green olivine. Basalt magma that slowly crystallizes underground becomes gabbro. This gabbro probably crystallized in a small magma chamber that fed volcanoes now completely lost to erosion.

From Waimea Canyon lookout, you can see three prominent volcanic formations: Most of the western wall of Waimea Canyon, including the rock at the lookout, is lava flows and hardened rubble of Nā Pali basalt; the lower wall of Waimea Canyon, directly across from Waimea Canyon lookout, is eroded in flows of Makaweli basalt; and the cliffs laced with waterfalls across the canyon are eroded in Olokele lavas.

Most of the countless lava flows in the canyon erupted from the summit caldera. Many of them drained down Makaweli graben on their way to the ocean. The canyon floor at this point roughly follows the western edge of the graben.

Toward the head of the canyon, you can see part of a buried fault, or slide scarp, separating thick flows of reddish Olokele basalt to the east from thin flows of brown Nā Pali basalt to the west. The Nā Pali flows dip down to the west.

Pu'u Ka Pele lookout is less than a mile north of Waimea Canyon lookout, and the view is similar.

The highest lookout, Pu'u Hinahina, perches above the deepest part of Waimea Canyon. The imposing lava cliffs of the Olokele Plateau make up the eastern wall in the upper reaches of the canyon. Flows of Nā Pali basalt make up the western wall. The canyon floor, thousands of feet below, roughly follows the western edge of the Makaweli graben.

From Pu'u Hinahina you can look across the water to the small islands of Ni'ihau and Lehua. On a clear day, you can see all the way to towering

The view toward the head of Waimea Canyon from Waimea Canyon Overlook is spectacular.

Looking down Waimea Canyon from Pu'u Hinahina Overlook.

Kaula Rock, 22 miles southwest of Niʻihau. This late-stage ash cone stands on a submerged shield volcano that was once a substantial island.

The Olokele Plateau

North of Puʻu Hinahina lookout, the highway winds through native upland rain forest to the edge of Olokele Plateau. To the north and left of the entrance to Kokeʻe State Park is a geophysical observatory. NASA scientists used it to track orbiting manned spacecraft before splashdown in the Pacific Ocean.

At the western tip of the Olokele Plateau, Hawaiʻi 550 reaches Kalalau lookout, and 0.8 mile beyond that, at the end of the road, Puʻu o Kila lookout. The overlooks provide views of Kalalau Valley, which reaches the Nā Pali shore 3 miles away and 4,000 feet below. Cliffs flanking the valley reveal thin flows of Nā Pali basalt that poured down the north flank of Waiʻaleʻale when it was still a growing shield volcano. Resistant dikes maintain the slender ridgelines. If you look carefully at the valley walls, you can see a few dikes cutting vertically through the basalt lava flows.

Alakaʻi Swamp, on the plateau behind the overlooks, is widely regarded as the wettest place on earth. Rainfall here is about 500 inches a year. Consider yourself lucky if the sun shines on you.

Kalalau Valley from Puʻu o Kila Overlook.

Glossary

'A'ā. A basalt lava flow with a rough, jagged surface.

Agglutinated. Melted together to form a single solid mass upon cooling.

Alkalis. The elements sodium and potassium.

Alkalic Basalt. Basalt enriched in the alkali element, sodium. Examples of alkalic basaltlike rocks include nephelinites, hawaiites, and ankaramites, among others.

Alluvial Fan. A fan-shaped deposit of stream sediment at the mouth of a canyon.

Alteration. Chemical change in a rock as a result of a reaction with hot water, steam, or volcanic gases.

Ankaramite. An alkalic basalt containing many large, black pyroxene crystals and a smaller number of green olivine crystals.

Ash. Fragments of lava fine enough to drift on the wind.

Ash Cone. A low, broad volcanic cone enclosing a wide, shallow crater. Volcanic ash makes up the rim of the cone.

Asthenosphere. A zone of soft, nearly molten rock in the earth's upper mantle, across which the plates slide.

Basalt. A common dark gray to black rock composed mainly of microscopic crystals of pyroxene and plagioclase. Some basalts also contain olivine and, in rare cases, other minerals.

Basanite. A variety of basalt containing tiny crystals of pale gray nepheline, and plagioclase.

Block Sag. A depression in an ash layer made by the weight of a block that dropped onto the ash.

Bomb. A large blob of lava thrown out by a volcanic explosion.

Bomb Sag. A depression in an ash layer made by the weight of a bomb that dropped onto the ash.

Breccia. Rock made up of angular, broken fragments contained in a fine-grained matrix.

Calcareous. Composed of calcite, entirely or in part.

Calcite. Calcium carbonate; the main mineral of coral reefs.

Caldera. A very large crater that formed as the surface collapsed into an emptying magma chamber. Calderas typically have a flat floor and steep sides.

Cinder. A bubbly clot of lava explosively ejected from a vent. Cinders typically range from the size of a pea to an orange.

Cinder Cone. A conical hill formed by the accumulation of cinder around a volcanic vent. Most cinder cones have craters in their summits.

Crossbeds. Layers within a bed of sand or ash that are inclined to the major or overall layers.

Debris Flow. A deposit of rock fragments and mud left by a sediment-laden flood.

Delta. A fan-shaped deposit of sediment left where a stream enters the sea. It also refers to a fan-shaped peninsula of land added where a lava flow descends a steep slope and enters the sea.

Dikes. Igneous intrusions that fill fissures, typically cutting through the enclosing rock at a high angle.

Domes. Mounds of volcanic rock that erupted in the form of lava too viscous to run out into a thin lava flow.

Dunite. A rock composed mainly of olivine and derived from deep in the earth's mantle.

Extrusion. The eruption of molten rock.

Fault. A crack in the earth's crust along which adjoining masses of rock slide.

Fault Scarp. A cliff or ledge marking the exposed surface of a fault.

Fissure. An open fracture.

Gabbro. A rock chemically identical to basalt and composed of the same minerals, pyroxene and plagioclase, in crystals large enough to be seen without a magnifier. Basalt magma that crystallizes below the surface becomes gabbro.

Graben. An area that dropped between parallel faults.

Hawaiite. A type of alkalic basalt.

Hot Spot. A volcanically active area of abnormally hot rock 50 to 150 miles across, which is underlain by a plume of hot rock rising from deep in the earth's mantle.

Ice Cap. A gigantic glacier capping a highland or mountain.

Intrusion. A body of intrusive igneous rock.

Intrusive Rock. Crystallized rock from a magma that did not erupt. Most intrusive rocks consist mainly of crystals large enough to be seen without a strong magnifier.

Isostasy. The flotation of the lithosphere on the asthenosphere.

Isostatic Sinking. The sinking of the lithosphere caused by placing a great weight upon it. The growth of Hawaiian volcanoes causes isostatic sinking.

Kīpuka. An area of older land surrounded by younger lava flows.

Late Stage of Volcanism. The period after the main, shield stage of Hawaiian volcano growth when eruptions become more explosive and less frequent. Late-stage volcanic rocks contain more alkalis than shield-stage rocks.

Laterite. A type of soil composed mainly of kaolin clay, aluminum oxide, and iron oxide, which stains it red. It is typical of regions with warm, moist climates and is extremely infertile. It supports sugar cane and pineapples.

Lava. Molten rock that flows out of the ground.

Lava Balls. Rounded boulders of smooth lava scattered across the surface of a lava flow. Many lie close to or inside flow channels.

Lava Pillows. Rounded masses of lava that form where lava pours out underwater.

Lava Shield. A gently sloping mound, typically no more than a few miles across, made up of thin lava flows around a vent.

Lava Tree. A hollow pillar of lava formed when a flow surrounds a tree trunk, cools hard around it, then recedes. The trunk burns, leaving behind the hollow lava pillar.

Lava Tube. A naturally formed tunnel in the center of a lava flow, commonly created by crusting over of the main lava channel, followed by drainage of the lava.

Lavacicles. Hardened drippings of lava from the ceilings of lava tubes and overhangs in a lava flow.

Levees. In lava flows, natural walls that build up along the edges of a flow.

Lithosphere. The outermost layer of material around earth, consisting of the crust and a slice of underlying mantle.

Lherzolite. A type of peridotite made up mainly of olivine, with somewhat lesser amounts of pyroxene and red garnet. Contains two varieties of pyroxene.

Magma. Molten rock beneath the earth's surface.

Magma Chamber. The central area of magma storage inside a volcano. Calderas typically form above magma chambers.

Mantle. The part of the earth between the core and the crust; the largest part of the earth. Peridotite is the main rock in the mantle.

Melilite. A rare mineral that crystallizes instead of feldspar in alkalic volcanic rocks extremely rich in sodium and deficient in silica. Melitite is virtually impossible to identify without the aid of a specialized petrographic microscope.

Moonstone. White feldspar with a bluish or pearly play of colors. Moonstone is polished as a gemstone.

Moraine. A mound of loose rubble piled up at the end of a glacier.

Mudflow. A mass of mud that pours down a slope. Mudflows are considerably denser than clear water and exert a correspondingly greater buoyant effect, which enables them to carry large rocks.

Mugearite. An alkalic variety of basalt in which the plagioclase feldspar is especially rich in sodium.

Nepheline. A mineral related to feldspar that crystallizes from alkalic magmas rich in sodium. It is hard to distinguish nepheline from quartz without the aid of a petrographic microscope.

Nephelinite. A strongly alkalic basaltlike rock containing the mineral nepheline instead of plagioclase feldspar.

Olivine. A light to dark green glassy mineral composed of iron, magnesium, and silica.

Outwash. Stream or flood deposits left by glacial meltwater.

Pāhoehoe. Basalt lava with a smooth, billowy, ropy surface.

Palagonite. Yellow clay that forms when glassy basalt reacts with water.

Pali. The Hawaiian word for cliff, precipice.

Paleosol. An ancient soil, buried by younger rocks.

Pele's Hair. Slender golden threads of basalt glass spun by falling droplets of lava or by lava blown out of a vent in an outrush of hot gases.

Peridotite. A dark rock composed mainly of olivine and pyroxene with minor amounts of other minerals.

Pillow Lava. A lava flow made of several rounded masses of basalt, formed when the lava enters water.

Pisolites. Small, rounded balls of ash, about the size of peas.

Plagioclase. A light gray to white variety of feldpsar, which typically forms blocky or needle-shaped crystals. Plagioclase contains sodium, calcium, aluminum, and silicon dioxide.

Plates. Large pieces of the earth's lithosphere—the crust and uppermost mantle.

Plunge Pool. A deep pool eroded at the base of a waterfall.

Potholes. Cylindrical holes eroded into bedrock in stream beds.

Pyroclastic Material. Ash, cinder, and other lava fragments blown out of a vent during a volcanic eruption.

Pyroxene. A group of minerals composed of iron, magnesium, calcium, and silicon dioxide in varying proportions. Most are dark green to black.

Pyroxenite. A rock composed mainly of pyroxene.

Rift Zone. A belt of fissures and volcanic vents stretching from the summit of a volcano to its base, which may be several miles away.

Rejuvenated Stage of Volcanism. A period of infrequent, explosive eruptions on Hawaiian volcanoes following a long interval of erosion. The last activity of a Hawaiian volcano.

Rock Glacier. A mixture of rock and ice, made up mostly of rock, that moves like a glacier.

Sea Stacks. Erosional remnants of rock standing in the surf.

Scarp. A cliff related to landsliding, slumping, or faulting.

Shield Stage of Volcanism. The period when a volcano is growing as a shield.

Shield Volcano. A broad, gently sloping volcano, typically tens of miles across, made mainly of thin lava flows.

Skylight. A natural opening in the roof of a lava tube.

Slide Scarp. A cliff created at the top of a landslide, where the slide tears loose.

Slump. A landslide.

Slump Scarp. A cliff created by a tear in the ground at the upper edge of a slump.

Spatter. Clots of thin, fluid lava thrown out of a vent, which splat as they hit the ground.

Spatter Cones. Small, rugged, steep-sided volcanic cones built from the accumulation and welding together of spatter.

Spatter Rampart. A low ridge of welded spatter next to a fissure vent.

Tholeiite. The most common type of basalt.

Till. Fine material with blocks mixed in, which is deposited by moving glaciers.

Trachyte. A pale gray volcanic rock composed mainly of plagioclase feldspar, with lesser amounts of pyroxene. It tends to erupt either explosively or in thick flows.

Tree Molds. Lava casts of tree trunks.

Trunk Stream. The main stream draining a valley, into which many tributary streams feed.

Tsunami. Giant waves raised by movement underwater, typically of a fault or landslide.

Tumuli. Steep mounds from a few feet to a few tens of feet across, in a pāhoehoe flow. They are heaved up by gases that accumulate inside a cooling flow.

Vent. The opening through which molten lava erupts.

Vesicles. Holes that form where gas bubbles are trapped in cooling lava.

Wave-Cut Bench (or **Platform**). A flat terrace of rock at tide level, carved by waves eroding the shore.

Weathering. The transformation of rock into soil.

Xenolith. A fragment of rock dragged up from below in rising magma.

Zeolites. A large family of pale minerals, some of which line the walls of vesicles in lava flows.

Additional Reading

Adams, W. M., and von Seggern, D. 1969. "Electronic mapping of Hawaiian lava tubes." *Association of Engineering Geologists Bulletin* 6 (2): 95-104.

Bier, J. A. 1988–1992. *Reference Maps of the Islands of Hawai'i.* Honolulu: University of Hawai'i Press. (Excellent color, topographic maps of Hawai'i, Maui, Moloka'i and Lāna'i, and Kaua'i.)

Bolt, B. A. 1993. *Earthquakes.* New York: Freeman and Co. 282 pp.

Brigham, W. T. 1909. The Volcanoes of Kilauea and Mauna Loa on the Island of Hawaii. Honolulu: Bishop Museum Press. 222 pp.

Clark, J. R. 1985. *Beaches of the Big Island.* Honolulu: University of Hawai'i Press. 204 pp.

———. 1985. *Beaches of O'ahu.* Honolulu: University of Hawai'i Press. 210 pp.

———. 1989. *Beaches of Maui County.* Honolulu: University of Hawai'i Press. 168 pp.

———. 1990. *Beaches of Kaua'i and Ni'ihau.* Honolulu: University of Hawai'i Press. 144 pp.

Decker, R. W., and Decker, B. 1989. *Volcanoes.* New York: Freeman and Co. 285 pp.

Decker, R. W., T. L. Wright, and P. D. Stauffer, eds. 1987. *Volcanism in Hawaii.* 2 vols. U.S. Geological Survey Professional Paper 1350. 2,506 pp.

Dibble, S. 1843. *History of the Sandwich Islands.* Lahaina, Maui: Lahainaluna Seminary. 65 pp.

Dudley, W. C., and M. Lee. 1988. *Tsunami!* Honolulu: University of Hawai'i Press. 517 pp.

Ellis, E. 1827. *Journal of William Ellis: Narrative of a Tour of Hawaii, with Remarks on the History, Traditions, Manners, Customs, and Language of the Inhabitants of the Sandwich Islands.* Reprinted 1963. Honolulu: Advertiser Publishing. 342 pp.

Fiske, R. S., and E. D. Jackson. 1972. "Orientation and Growth of Hawaiian Volcanic Rifts: The Effect of Regional Structure and Gravitational Stresses." *Proceedings of the Royal Society of London* A329:299–326.

Hazlett, R. W. 1987. *Geological Field Guide, Kilauea Volcano.* Volcano, Hawai'i: Hawaiian Volcano Observatory and the Hawaii Natural History Association. 121 pp.

Heliker, C. 1993. *Volcanic and Seismic Hazards on the Island of Hawaii.* Honolulu: Bishop Museum Press. 52 pp.

Heliker, C., and D. Weisel. 1990. *Kīlauea, The Newest Land on Earth.* Honolulu: Bishop Museum Press. 76 pp.

Hitchcock, C. H. 1911. *Hawaii and Its Volcanoes.* Honolulu: The Hawaiian Gazette. 314 pp.

Macdonald, G. A., A. T. Abbott, and F. L. Peterson. 1983. *Volcanoes in the Sea: The Geology of Hawaii.* 2nd. ed. Honolulu: University of Hawai'i Press. 517 pp.

Mattox, S. R. 1994. *A Teacher's Guide to the Geology of Hawaii Volcanoes National Park.* Volcano, Hawai'i: Hawaii Natural History Association. 392 pp.

Moore, J. G., W. B. Bryan, and K. R. Ludwig. 1994. "Chaotic Deposition by a Giant Wave, Molokai, Hawaii." *Geological Society of America* 106:962–967.

Moore, J. G., et al. 1989. "Prodigious Submarine Landslides on the Hawaiian Ridge." *Geophysical Research* 94:17465–17484.

Moores, E. M., ed. 1990. *Shaping the Earth; Tectonics of Continents and Oceans: Readings from Scientific American Magazine.* New York: Freeman and Co. 206 pp.

Stearns, H. T. 1983. *Memoirs of a Geologist: From Poverty Peak to Piggery Gulch.* Honolulu: Hawaii Institute of Geophysics. 242 pp.

Wright, T. L., D. L. Peck, and H. R. Shaw. 1976. "Kilauea Lava Lakes: Natural Laboratories for Study of Cooling, Crystallization, and Differentiation of Basaltic Magma." In G. H. Sutton, N. H. Managhnani, and R. Moberly, eds., *The Geophysics of the Pacific Ocean Basin and Its Margin.* Washington, D.C.: American Geophysical Union, 375–390.

Index

About the Authors

Richard W. Hazlett
is a geology professor at
Pomona College in Claremont,
California. He has been an
interpretive ranger at Hawaii
Volcanoes National Park,
a research technician at the
Hawai'i Volcano Observatory,
a geology professor at the
University of Hawai'i at Hilo,
and a lead scholar for the
24-part PBS television series
"Earth Revealed."

Donald W. Hyndman
is cofounder of the
Roadside Geology series,
books that bring geology
to the general public. When
he is not writing or editing
books, Hyndman teaches
geology at the University
of Montana in Missoula.